Fundamentals of Organic Neuromorphic Systems

Victor Erokhin

Fundamentals of Organic Neuromorphic Systems

 Springer

Victor Erokhin
Italian National Research Council
Institute of Materials for Electronics
and Magnetism
Parma, Italy

ISBN 978-3-030-79494-1   ISBN 978-3-030-79492-7   (eBook)
https://doi.org/10.1007/978-3-030-79492-7

This Springer imprint is published by the registered company Springer Nature Switzerland AG
The registered company address is: Gewerbestrasse 11, 6330 Cham, Switzerland

Preface

Dear reader,

In this book I have tried to summarize more than 15 years of activity in the field of organic neuromorphic systems. You will find different aspects of neuromorphic systems, including fabrication, properties, characterizing techniques, and possible applications. I tried to make a comprehensive overview of fundamental principles and technical approaches for the implementation of neuromorphic systems with perspectives of their further development, and I hope it will be useful for a wide range of readers, from MS and PhD students to experts, working in complimentary fields, including physicists, chemists, engineers, biologists, physicians, and (I hope I was rather clear) even humanitarians.

Parma, Italy   Victor Erokhin

Acknowledgments

First of all, I want to thank my wife Svetlana for her help in the preparation of this book. She was involved in several of my original research works, described in the book, and helped me in preparation of figures. I am also very grateful to Svetlana, who, starting from the university period, was always the harsh critic of all my ideas and manuscripts and helped me to improve this particular book.

I want to express my special thanks to Prof. Lev A. Feigin. His help and advice, since the time of my PhD work under his supervision, were very useful for me in all aspects of scientific and social life.

My special thanks to Dr. Tatiana Berzina and Prof. Marco P. Fontana (University of Parma), with whom I have published my first works on this book's subject. During this period, an essential contribution was made by Dr. Anteo Smerieri, who completed his MS and PhD theses works with me. In particular, Sect. 4.1 is a part of his master's thesis work, while Sect. 6.3 is a part of his doctoral work. Regarding members of the University of Parma, I want to thank Dr. Paolo Camorani and D. Maura Pavesi, as well as Dr. Valentina Allodi, Dr. Konstantin Gorshkov, and Dr. Francesca Pincella, who did their MS or PhD works in our group.

I want to express special thanks to chemists from the University of Pisa, involved in works on the synthesis of materials: Dr. Giacomo Ruggeri, Dr. Andrea Pucci, Dr. Lucia Ricci, and Marco Bernabò.

I want to thank a lot Dr. Oleg Konovalov from ESRF (European Synchrotron Radiation Facilities). Important step in the understanding of the working principles of organic memristive devices was taken during experiments carried out in his station. Moreover, discussions with him are not only useful, but also pleasant for me.

Our collaboration with scientists from Max Planck Institute for Biological Cybernetics was very fruitful. In particular, Prof. Almut Schüz deepened my understanding of brain architecture and roles of its different parts. Dr. Rodrigo Sigala has applied mathematical algorithms, developed for brain function description, to artificial systems. It was my great pleasure to collaborate with Prof. Valentino Breitenberg, whose constructive comments were extremely useful for linking properties of artificial systems with natural ones.

Working on the artificial reproduction of the part of the pond snail nervous system, our long-drawn discussions with Prof. Jianfeng Fend and Dr. Dimitris Vavoulis from the University of Warwick were very useful.

I express a special thanks to Prof. Andrew Adamatzky from the University of West England, who is always ready to discuss new, even strange, ideas in the field of unconventional computing. I am also grateful to Dr. Ella Gale and Dr. Gerard David Howard for useful discussion.

Many thanks to colleagues from the Institute of Materials for Electronics and Magnetism, Italian National Research Council (IMEM, CNR), where I am working now. First of all, I want to thank Dr. Silvia Battistoni, bright young scientist who did her PhD work in our group and continues producing high-level research papers, and Dr. Salvatore Iannotta, with whom we have prepared and carried out several successful projects at international and national levels. Of course, I want to thank all my colleagues at IMEM-CNR for supporting our group's activity (Smart and Neuromorphic Biointerfacing Systems) and providing a friendly atmosphere at the institute, stimulating our research activity. I also want to thank my colleagues from Torino for the technological support: Prof. Matteo Cocuzza, Dr. Simone L. Marasso, and Dr. Alessio Verna, working in tight collaboration with IMEM-CNR.

In the past few years, many results described in this book were obtained during collaboration with the National Research Center "Kurchatov Institute." I am very grateful to Dr. Vyacheslav Demin and Dr. Andrey Emelyanov for very useful discussions that resulted in planning and carrying out of new experiments, improving the understanding of neuromorphic systems. I want to also thank Dr. Yulia Malakhova, who took the responsibility of not only carrying out research activity, but also coordination of the administrative aspects of our collaboration. Important contributions were also made by two bright young students: Dmitry Lapkin and Nikita Prudnikov. I am also very grateful to the following people: Prof. Mikhail Kovalchuk, Prof. Sergey Chvalun, Prof. Pavel Kashkarov, Dr. Anton Minnnekhanov, Dr. Vladimir Rylkov, and Dr. Alexey Korovin, with whom I have discussed different aspects of neuromorphic systems.

The part related to the coupling of artificial synapses with live neurons would not have been completed without the contribution of three people from Kazan Federal University: Prof. Rustem Khazipov, with whom we have discussed and planned experiments; Dr. Marat Mukhtarov, who organized the experimental setup and data acquisition protocols, as well as took an essential part in the discussion of the data interpretation; and Elvira Juzekaeva, whose "green fingers" provided the connection of artificial system with the natural one. I also want to thank other colleagues from Kazan Federal University: Dr. Max Talanov, Prof. Igor Lavrov, and Prof. Evgeny Zykov for useful discussions on the possible neuroprosthesis.

In addition, I am very grateful to Prof. Leon Chua, Dr. Jullie Grolier, Dr. Sandro Carrara, and Dr. Alexey Mikhailov for the very useful discussions I had with them.

Last but not least, I want to express my warmest thanks to my friend Attilio Anselmo, who is a real fan of science and was my neighbor in the period of 1994–2002. Our weekly discussions helped me formulate my statements better.

Introduction

Bio-inspired information processing systems began attracting the attention of researchers since the 1950s [1–3]. Initially, only hardware realizations of neural networks were considered. Later, the explosive development of traditional computers, whose elementary basis and information processing principles were significantly different from those used by ordinary people, resulted in a situation where main efforts towards the realization of bio-inspired (neural) system were shifted to the direction of their implementation at the software level [4–6].

This decrease in activity, connected to the hardware realization of bio-inspired systems, was due to the lesser capabilities and slower development of the microelectronics technology with respect to the intensive development of traditional computers. It became much easier to implement properties of neurons and synapses by appropriate software than to fabricate physical devices with required properties.

Nevertheless, this task is still very real as the hardware realization of bio-inspired (neural) systems is expected to allow a significant increase in computational power, a decrease of energy consumption, and a better understanding of the working principles of the nervous system and brain, providing the necessary model elements and systems as well as allowing experiments that are impossible to perform directly on animals and humans. A very important feature of such systems will be the capability of parallel information processing, when many input signals will be treated simultaneously, which is now impossible with traditional single-nucleus computers: they make only one operation in a fixed moment.

Successful realization of bio-inspired systems requires the use of electronic compounds, having several specific properties. Some of the properties, necessary for the realization of electronic circuits, mimicking functions of nervous system and brain, are listed below:

- Involvement of same elements for memorizing and processing of the information
- Variation of electrical properties according to the Hebbian (or alternative) learning rule (electronic synapse) [7]
- Possibility of working in the auto-oscillation generation mode
- Formation of stable chains of the signal transfer

- Materials used for electronic compounds must allow self-organisation into 3D systems mimicking intrinsic brain functions

Each of the listed properties is necessary for the realization of bio-inspired systems. Let us consider them.

Involvement of Same Elements for Memorizing and Processing of Information

This property is the basis of the fundamental difference in the architecture of the nervous system and brain with respect to modern computers. In computers, memory and processor are separate independent systems, having no influence on each other. In this case, information plays a passive role. It can be recorded, accessed, and canceled. However, processor properties are not changed in this case. In the case of the nervous system, the situation is absolutely different: the same elements are used for memorizing and processing of information. Thus, information plays an active role: it is not only memorized, but it changes connections within the processor, which allows learning at the "hardware" level. Variations of mutual connections of elements will simplify resolving of similar tasks in the future. Electronic elements, mentioned above, must provide similar tasks in bio-inspired electronic circuits.

Variation of Electrical Properties According to the Hebbian (or Alternative) Rule (Electronic Synapse)

Hebbian rule [7] is still considered as the main algorithm responsible for the variation of synaptic connections in the nervous system and learning of living beings. In its classic formulation, the Hebbian rule claims:

- *"When an axon of cell A is near enough to excite cell B and repeatedly or persistently takes part in firing it, some growth process or metabolic change takes place in one or both cells such that A's efficiency, as one of the cells firing B, is increased"*

In terms of electronic circuits, the rule can be rewritten in the following way. Cells A and B represent nonlinear threshold elements, activation (spike generation) of which occurs only when the integral of the signal, coming to the element in a certain interval of time, overcomes a determined threshold level. These elements are connected through a contact, the resistance of which depends on the duration/frequency of its involvement into the process of signal transfer from element A to element B. Thus, as longer or more frequently the contact is used, there is greater probability that element B will be activated after the appearance of activity on element A. The resistance of the contact, therefore, is the function of the history of its involvement into the formation of the signal transfer chains.

Currently, the logic modification of this rule, called Spike Timing Dependent Plasticity (STDP) [8–13], is more frequently used for the description of learning. It should be noted that the propagation of signals in the form of spikes is a necessary requirement for living beings, where all processes are of electrochemical nature. Thus, if the information transfer processes would occur in DC mode, we would have a directed transport of ions resulting in the appearance of gradients of elements distribution concentrations and electric fields, which will stop the nervous system from functioning. In the case of electronic circuits, generally speaking, the use of pulses is not required, as electrons will carry the information. Therefore, for analog circuits we can apply the Hebbian rule in its classic formulation, presented above. However, it seems that STDP mechanism is the only one explaining unsupervised learning, establishing automatically causal relationships, which will be considered in the appropriate part of the book.

Possibility of Working in Auto-Oscillation Mode

The possibility of auto-oscillations generation during fixed configuration of input stimuli is also a very important property for the realization of bio-inspired systems. According to the statement of Schrödinger, "Living matter evades the decay to equilibrium" [14]. It means that internal rhythmic processes must occur in a system even when environment is not changed. In the context of computer science, these processes can play the role of clock generators. However, in contrary to computers, this frequency will not be fixed and can be varied according to the state of the system in each moment and configuration of external stimuli. In addition, the presence of such processes seems to be an essential requirement for the appearance of creative processes: overlay of the external stimuli onto the internal activity of the system will result into the formation of new connections in the network, which would result in the appearance of new unexpected associations.

Formation of Stable Chains of the Signal Transfer

Bio-inspired neuromorphic system must provide a balance between two important requirements. On the one hand, the realized system must have plasticity: it should allow adaptations and learning as a function of the presence of external stimuli, internal activity, and environment variation. On the other hand, it must have individual properties, determining its own behavior in each particular situation. Thus, it requires the establishing of stable long-term connections, which are formed mainly at the early stages of the system learning. In analogy with animals and humans, it can be connected to so-called "baby learning" or imprinting [15–17]. At this stage, stable chains of the signal transfer are formed, which determines the character of a particular individual (or system in the case of artificial networks).

Materials Used for Electronic Compounds Must Allow Self-Organisation into 3D Systems Mimicking Intrinsic Brain Functions

This requirement is also related to the fundamental difference of brain and computer architectures. Defined determined architecture of the computer is based mainly on the capabilities of the current state of electronic technology. This technology, CMOS-based one, is mainly planar, where all electronic elements are arranged in the same layer. Current state of the technology allows multilayer organisation of active components. However, the number of layers is strongly limited. Instead, the nervous system and brain have 3D organization with the possibility of connections even between rather distant threshold elements (neurons). Moreover, even if there are similarities of brain architecture among humans and different classes of animals, each individual has its own particularities in the organization of nervous system and brain, in which connections can also be varied during individual learning processes. Therefore, if we really want to realize a bio-mimicking neuromorphic system, we should avoid using inorganic materials, because only organic molecules allow self-organization in 3D systems with distant connections between threshold elements, allowing the variation of the strengths of connections due to the external stimuli and internal activity.

Realization of elements with properties, listed above, will allow to build not only computational systems of a new type, but will also allow to better understand the brain function mechanisms and to make individual and group behavior models as a result of learning, internal processes, and presence of external stimuli. The following factors will affect the behavior:

- Fabrication technology. This factor is similar to the genetic code, responsible for the particularities of the organization of nervous system of classes and types of living beings (the organization is determined by the genome of this particular type).
- Early stage learning (imprinting). At this stage, the individual properties of the brain are formed, which determine the character of a particular individual (not the type in general). The depth of imprinting depends strongly on its correspondence to genetic factors (in the case of artificial systems, applied training algorithms must take into account the architecture of the system and technological approaches, used for its fabrication). These induced particularities remain practically not variable during the whole life of the individual.
- Everyday learning. Adaptation of the behavior is due to the variations of the environment and external stimuli, which are superimposed on the system with acquired experience and internal on-going non-linear processes, slightly modifying connections in the brain.

Last year's special electronic elements, memristors of memristive devices, are considered as good candidates for performing specific functions, listed above. These devices have hysteresis loop in their cyclic voltage-current characteristics and, therefore, perform memory effects. This fact was the reason for their initial consideration as elements for the realization of non-volatile memory units. However, currently a wider range of areas of their possible applications is considered.

Most of the realized memristive devices are fabricated from inorganic materials, mainly metal oxides. However, a wide range of other materials, used for the memristive devices fabrication, has been reported.

Organic polyaniline-based device was developed directly for mimicking such synapse properties as the presence of memory (hysteresis loop of voltage-current characteristics) and rectification (unidirectional signal propagation).

In this book, we will present an overview of different kinds of memristive devices, with special attention to the organic polyaniline-based systems, because, up to now, only these elements satisfy all five requirements listed above.

Contents

About the Author

Victor Erokhin received his MS degree in physics and engineering in 1983 in Moscow Institute of Physics and Technology and PhD in physical and mathematical sciences in 1990 in the Institute of Crystallography, Russian Academy of Sciences. After his MS degree (1983–1987), he worked as an engineer in the applied research institute "Delta" (Moscow, Russia). In the period 1990–1992 (after PhD thesis), he was a researcher in the Institute of Crystallography, Russian Academy of Sciences. In 1992, he was invited to work in Genoa University (Italy), and during the period 1992–2003, he worked in different industry-oriented companies, nucleated around Genoa University (initially, senior scientist, later head of research units). After the Genoa period, during 2003–2011, he was a leading scientist in INFM (National Institute of Physics of Matter) and visiting professor of Parma University (Italy). Since 2011, he has worked in the Institute of Materials for Electronics and Magnetism, Italian National Research Council (Parma, Italy). His current position is director of research and head of the research unit "Smart and Neuromorphic Biointerfacing Systems" in the same institute. Victor Erokhin is an author of more than 200 research papers in international refereed scientific journals, 19 book chapters, and 13 patents. He is editor-in-chief of BioNanoScience (Springer Nature) and a member of the editorial board: International Journal of Parallel, Emergent and Distributed Systems, Electronics. Victor Erokhin was the principal investigator in numerous national and international research projects, the chairman and cochairman of International Symposia, and a member of numerous national and international committees, including evaluation panels of European Synchrotron Radiation Facilities (ESRF, Grenoble, France) and Nano-Tera Projects (Switzerland).

Chapter 1
Memristive Devices and Circuits

1.1 Determination of Memristor

The meaning of the word "memristor" is a resistor with memory. Ideally, the resistance of the device must be a function of the integral of the current that has passed through it.

The concept of "memristor" was introduced in 1971 by Leon Chua [18]. Considering the symmetry of electronic circuits, he suggested that there is a missed passive electronic element that must connect a charge with magnetic flux. Later, the concept of "memristor" was expanded and elements with memory began to be called "memristive devices and systems" [19]. Currently, there are efforts to expand even more the class of devices which can be considered as memristors [20]. The authors discuss four types of systems that can be considered as memristors.

According to such classification, practically all systems with memory, even living beings, can be attributed to one of the four mentioned classes of memristors. Of course, such classification seems to be rather artificial. Moreover, the memristor in its classic definition of L. Chua cannot exist [21–23]. However, the concept and term have done a great impact on successive works in the field of resistive-switching devices and systems. Therefore, it seems useful to present here the main steps of this approach.

Initial idea can be illustrated by Fig. 1.1.

In his considerations, L. Chua considered four fundamental parameters of electronic circuits: Voltage, current, charge, and magnetic field flux. Pairs of these parameters are connected by simple equations. For example, the current is a time derivative of charge. For three cases there are equations, linearly connecting these parameters and three coefficients correspond to properties that are attributed to three known passive elements. For example, the proportional coefficient between the variation of the voltage and current is the resistance and the device, having this property, is called a resistor. Capacitors connect the variation of the voltage with the

© Springer Nature Switzerland AG 2022
V. Erokhin, *Fundamentals of Organic Neuromorphic Systems*,
https://doi.org/10.1007/978-3-030-79492-7_1

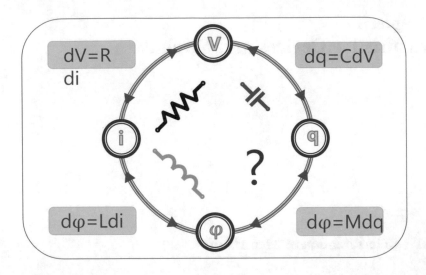

Fig. 1.1 Schematic representation of the position of "memristor" among other three fundamental elements of electronic circuits

variation of the charge and inductance that connect the variation of the magnetic flux with the variation of the current.

According to the hypothesis of L. Chua, there is a missing element that must connect the charge and magnetic field variations:

$$d\varphi = Mdq \qquad (1.1)$$

By simple mathematical calculations, it has been shown that if memristance M is independent of the charge variation; Eq. 1.1 is simply Ohm low and M represents only the resistance of the element. However, if M is a function of charge q, the relationship between the voltage at the device terminals and the charge, passed through it, is determined by the following equation:

$$v(t) = M(q)i(t) - M\left(\int_{-\infty}^{t} i(\tau)d\tau\right)i(t) \qquad (1.2)$$

At each particular moment, the behavior of the memristor is similar to that of the resistor with the actual value of the resistance, determined by the temporal integral of the current, passed through it. Thus, the memristor is the resistor with memory with the actual value of the resistance, determined by the history of its previous functioning.

Later, L. Chua has claimed the criterium of "passivity" for memristive systems to distinguish them from the dynamic system: the lack of the capability to accumulate energy, similar to capacitor and inductance; lack of the phase shift between input and output signals (it means that voltage-current characteristics always path through zero point; this statement was shown to be wrong for real devices that will be discussed later); at high frequencies, the behavior of memristive systems is similar to that of resistors.

It is to note that all existing elements have resistance, capacity, and inductance (even if relative values of these parameters are different). For this reason, it seems logic to suppose the existence of "memcapacitors" and "meminductances." Basing on this suggestion, Y. Pershin and M. Diventra have generalized the statement [24], which was confirmed by available data on memcapacitors [25–36] and meminductors [25, 26, 37–40].

It is to note that even before L. Chua's work, an element with memory was suggested by B. Widrow [41] and called it "memistor." Similar to a transistor, the device was a three-terminal element. However, the principal difference between the transistor and memistor is that in the case of transistor the value of the current between source and drain electrodes is a function of the voltage applied to the gate electrode, while in the case of memistor this value is a function of the charge passed in the circuit of the gate electrode.

The first working memistors were made of pencil leads immersed in test tubes containing sulfate-sulfuric acid plating baths. The resistance of these elements was varied in the range of 100–1 Ohm with a typical time of 10 seconds.

It is to note, that it was theoretically demonstrated that memistors can be used for the simulation of plasticity of synapses [42].

Despite the fact that this element was considered as a very perspective one for the realization of artificial neural networks, it did not find wide applications. Initially, it was due to the fact that it was based on the galvanic cell and not on solid materials; later, main efforts in the direction of artificial neuron networks were concentrated in their software implementation.

1.2 Mnemotrix

The term "mnemotrix" was introduced by Valentino Braitenberg in his mental experiment on vehicles [43]. These vehicles were equipped with sensors, whose signals controlled the vehicle motion. A different number of sensors and their connections with motors determined one or other type of vehicle behavior. Mnemotrix is an element allowing "learning" of the vehicle through associations.

According to the V. Braitenberg definition, "...we buy a role of special wire, called Mnemotrix, which has the following interesting property: its resistance is at first very high and stays high unless the two components that it connects are at the same time traversed by an electric current. When this happens, the resistance of Mnemotrix decreases and remains low for a while...."

As it is possible to note, this element is rather similar to the device that could be used in electronic circuits allowing learning according to the Hebbian rule [7].

The approach suggested by V. Braitenberg is used currently in robotics [44].

It is to note that mnemotrix was considered as a benchmark for the realization of the organic memristive device – the main subject of the present book. In fact, before 2008, practically, nobody knew the term "memristor," as only four works have been published till this year: two of them were by L.O. Chua [18, 19] and two others from other researches [45, 46]. As the first work on this device was published in 2005 [47], it was called "electrochemical element for adaptive networks" and we began to call it "organic memristive device" only after 2008.

1.3 First Mention About the Experimental Realization of Memristor

As it was mentioned above, before 2008, only a few works on memristors were published, and all of them were theoretical ones.

The situation has been radically changed in 2008 when the research group from Hewlett Packard published in Nature the article with a very ambitious title "The missing memristor found" [48]. Immediately, after the publishing of this work, the activity in the field was explosively increased. According to the ISI Web of Science database, more than 6000 papers were published in the period 2008–2019, and their number still continues to increase.

Memristor of Hewlett Packard was composed of a thin film of titanium dioxide (about 50 nm) between two platinum electrodes. The scheme, illustrating the working principle of the device is shown in Fig. 1.2.

The thin film of TiO_2 can be schematically divided into two zones, one of which with a high concentration of oxygen vacancies had much lower resistance with respect to the other zone of pure titanium dioxide (in reality, there is not a strict boundary between these zones, but the gradient of resistance, as is shown in

Fig. 1.2 Schematic representation of the HP memristor (up), its equivalent circuit (middle), and linear distribution of the resistivity on the length of the active zone (bottom)

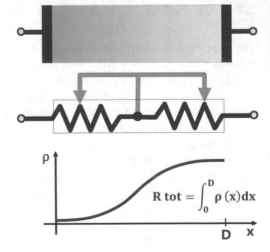

$$R \, tot = \int_0^D \rho \, (x) dx$$

Fig. 1.2). It was suggested that oxygen vacancies, having the electric charge, can move in the electric field, resulted from the voltage application, varying, therefore, the relative input of high and low resistance zones in the total conductivity of the element, as is shown in the equivalent electric circuit in Fig. 1.2.

Variable w of the device state, responsible for the memristor behavior, is the position of the boundary between zones with high and low resistances. When the voltage $v(t)$ is applied to metal electrodes, oxygen vacancies move in the titanium dioxide semiconductor, which results in the redistribution of the relative ratio of the contribution of these zones into the total resistance of the device. Therefore, the element can be in a high conducting state when the oxygen vacancies are rather homogeneously distributed through the whole titanium dioxide layer, or in a low conducting state when oxygen vacancies are shifted to one of the metal electrodes.

For the simplest case of the Ohmic conductivity and linear motion of ions in homogeneous mediumwith ion mobility μ_v, the behavior can be described by Eq. 1.3:

$$v(t) = \left(R_{ON}\frac{w(t)}{D} + R_{OFF}\left(1 - \frac{w(t)}{D}\right) \right)i(t) \qquad (1.3)$$

where i is a current, passed through the film, R_{ON} is the resistance of the highly conducting zone, R_{OFF} is the resistance of the low conducting zone, and D is the total thickness of the film.

The variation of the device state variable is determined by Eq. 1.4:

$$\frac{dw(t)}{dt} = \mu_v \frac{R_{ON}}{D} i(t) \qquad (1.4)$$

what results in Eq. 1.5 for the $w(t)$:

$$w(t) = \mu_v \frac{R_{ON}}{D} q(t) \qquad (1.5)$$

Inserting the value of 1.5 into Eq. 1.3 we will have the equation, describing the memristance of the system that in the case of $R_{ON} \leq R_{OFF}$ can be simplified to 1.6:

$$M(q) = R_{OFF}\left(1 - \frac{\mu_v R_{ON}}{D^2}q(t) \right) \qquad (1.6)$$

where the q-dependent term in the right part of Eq. 1.6 gives the main input to the memristance, and it increases in absolute value for higher mobilities of oxygen vacancies μ_v and lower thickness of the semiconductor layer.

The presence of the factor $1/D^2$ in the equation results in the fact, that the observed phenomenon is 10^6 higher for devices with nanometer sizes with respect to those with micron sizes. Authors of [48] used this fact for explaining why the memristor was not found before this work: technological capabilities did not allow the realization of required thicknesses of the active layer.

It is to note that the paper [48] was not the first one dedicated to such kinds of devices. Several works were published before. However, authors of [48] were the first ones, connecting such kinds of devices with the term "memristor." Before [48] "resistive switching memory" was more frequently used for these elements [49–60]. Nevertheless, currently, the terms "memristor" and "memristive device" are widely used for the elements, varying their resistance during functioning. The term "memristive device" is used more often due to the fact that the classic memristor of L. Chua is very likely cannot be realized due to principal considerations [21–23].

1.4   Inorganic Memristive Devices

Most inorganic memristive devices have a structure similar to that, considered in the previous section: a thin layer of metal oxide is placed between metal electrodes [61–83], and at least one of them is done from platinum.

It is to note that not only titanium oxide was used for the formation of active layers, but also oxides of other metals, such as, for example, Al_2O_3 [84–88], Gd_2O_3 [89], VO_2 [90–93], and others. In the last years, HfO_2 is widely used for the realization of memristive devices that is connected to the fact that it simplifies the production of such devices, because this material is used in existing electronic technology [94–103]. Another currently popular material for memristor production is graphene [104–117] that is very likely due to the unique structure and properties of this material.

Currently, the mechanism of the formation of the conducting filaments inside the insulating layer with their successive variation of the conductivity due to redox reaction is the most common for explaining the resistance switching [81, 100, 118–137].

This mechanism of the resistance switching in most of the inorganic materials implies the formation of the initial filament forming process what is done by the application of a rather high voltage between metal electrodes. The value of this voltage is much higher than the range in which the devices successively operate.

The electroforming process is schematically illustrated in Fig. 1.3 [88].

Voltage-current characteristics of devices made from different materials can have some differences with respect to that shown in Fig. 1.3 (current values at certain voltages, switching voltage), but the essential features of all of them are similar to the shown one.

Let us consider the characteristic shown in Fig. 1.3. Initially, the device is in a low-conducting state. The switching of the conductivity occurs at about 7.5 V. As it is clear from the figure, a sharp increase of the conductivity takes place at this point. A horizontal line of the current value after this voltage is due to the inserted value of the maximum possible current value (in order to protect the measuring unit from

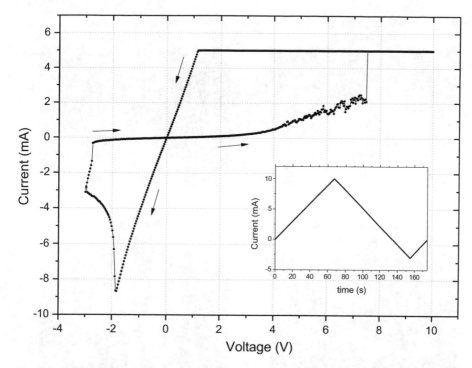

Fig. 1.3 Cyclic voltage-current characteristics of the memristive device based on Pt/Al$_2$O$_3$/Ti structure. Arrows indicate the direction of the applied voltage variation. The temporal dependence of the voltage variation is show in the inset [88]. (Reproduced with permission from Baldi et al. [88]. Copyright (2014) IOP Publishing, Ltd.)

damage. After the passing of the point, corresponding to the applied voltage of about +7.5 V, the conducting filament is formed in the insulating Al$_2$O$_3$ layer.

It is to note that such mechanism of randomly formed conducting filaments implies the main disadvantage of metal oxide-based memristive devices: the value of the voltage at which the resistance switching takes place can vary (even for several hundreds of millivolts) not only from one device to the other but also for different cycles of the same device.

It is also to note that the ideal memristor, according to the definition of L. Chua, must have a zero-current value at zero applied voltage. However, this requirement is not valuable for a lot of resistance switching devices.

It was developed a model, involving redox reactions and implying the appearance of electromotive forces, for the explanation of this behavior [138].

It was identified at least three reasons, which can result in such behavior: Nernst potential, diffusive potential, and Gibbs-Thompson potential. The contribution of each of the mentioned potentials to the appearance of the electromotive force is illustrated in Fig. 1.4 [138].

The described behavior (not zero current at zero applied voltage) has been registered experimentally [138]. The observed results are shown in Fig. 1.5.

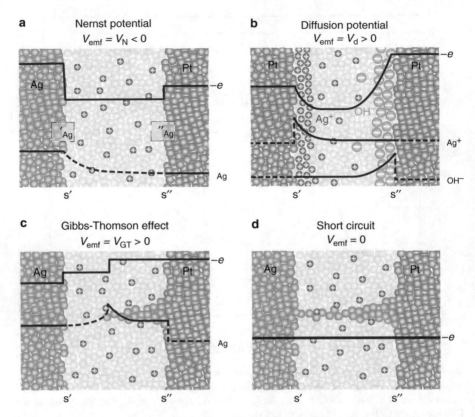

Fig. 1.4 The appearance of electromotive forces in memristive devices. (**a**) Nernst potential appears due to the difference of chemical potentials of metallic silver at silver/electrolyte and platinum/electrolyte interfaces. (**b**) Diffusion potential appears in the Pt/SiO$_2$/Pt cell due to the gradient of chemical potentials of Ag$^+$ and OH$^-$ ions; (**c**) In the case of filament formation, a gradient of the chemical potential (Gibbs-Thompson potential) appears in the structure; (**d**) The value of the electromotive force becomes zero when the formed conducting filament results in the shorting of metal electrodes. (Reprinted from: Valov et al. [138])

As it is clear from Fig. 1.5, when the device is in a low-conducting state, the voltage-current characteristics do not pass through the origin of the coordinates point. Instead, for a high conducting state it does.

The presented considerations show the importance of the redox reactions consideration even for memristive devices, based on inorganic materials (for organic materials it will be shown in appropriate sections). It also demonstrates the possibility of the appearance of nonzero current when zero voltage is applied.

Memristor devices, based on ferroelectric materials are a separate class of such elements [139–153]. In particular, in [154] it was demonstrated that the ON/OFF ratio of about 100 can be reached within 10 ns time due to the voltage-controlled variation of the tunneling barriers between ferroelectric domains.

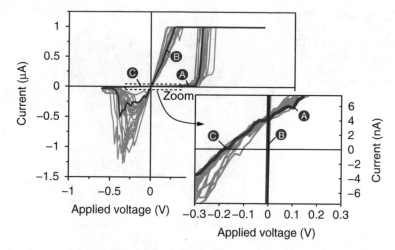

Fig. 1.5 Cyclic voltage-current characteristics, measure in the structure Cu/SiO$_2$/Pt (red curve – medium values; gray curves – statistics of experimental data) [135]. Points A and C correspond to a low-conducting state of the device, point B corresponds to a high-conducting state of the device. (Reprinted from: Valov et al. [138])

The main applications of memristive devices based on inorganic materials are memory and logic elements.

However, there are also other areas of applications of these elements, three of that seem to be very interesting.

A system, mimicking the learning of Pavlov's dog, in particular, the association of the neutral signal (bell ring) with the presence of food was realized in [155]. It is to note that a similar system was realized before on organic memristive devices [156]. This system was done for mimicking learning of the pond snail *Lymnaea stagnalis*, imitating not only learning but also the architecture of the part of the nervous system, responsible for the performing of this function. The scheme has been reconstructed by analyzing the available data on the signal propagation, obtained from a system of microelectrodes, implanted into the animal [157–163]. A detailed description of the obtained results will be presented in the appropriate section. Here we describe the approach, used in [155] for the system, based on inorganic devices.

The circuit shown in Fig. 1.6 was realized for mimicking such kind of learning.

As it is clear from Fig. 1.6, the circuit contains two inputs: one of them (UCS) corresponds to the presence of food, while the other one (CS) corresponds to the bell ring. Learning takes place only when both signals are applied. After this event, the synaptic strength (IN) is reinforced to such a degree that successive application of the CS input only results in the fact that the passed signal value is enough for firing the motor neuron (M). The approach is rather similar to that used for mimicking the pond snail learning [156].

It was enough to use one memristive device only for the realization of the circuit [155]. Obviously, the circuit cannot pretend to mimic the architecture of any part

Fig. 1.6 The scheme of the circuit, used for mimicking Pavlov's dog learning (UCS unconditional stimulus, CS conditional stimulus, IN strengthen interneuron, M motor neuron)

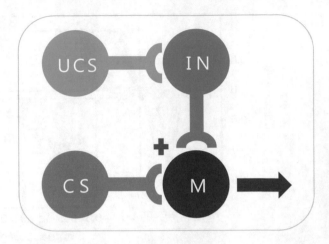

Table 1.1 Experimental results of the circuit mimicking Pavlov's dog learning

UCS (food)	CS (bell)	Output signal (before learning)	Output signal (after learning)
1	0	1	1
1	1	1	1
0	**1**	**0**	**1**
0	0	0	0

(even very simplified) of the nervous system of a dog. It is to note that it was necessary to use two memristive devices for imitating the learning of a more simple animal (pond snail), and their position was practically the same, as in the model of the part of the nervous system, responsible for such learning. However, even a very simple design of the circuit, suggested by authors [155], allowed to obtain the desired result: after learning (simultaneous application of both stimuli) the application of the initially neutral stimulus results in the significant increase of the output signal. In addition, the system contains the threshold element that starting of the function begins only when the value of the output signal is higher than a certain value.

The results of the experiments on learning of the system [155] are presented in Table 1.1.

As it is clear from Table 1.1, learning corresponds to the last but one row: before learning the signal, corresponding to the bell ring does not result in the appearance of the output signal, while after learning it does.

It is to note that the configuration of output signals after learning is linearly separable, which can be realized by a single layer elementary perceptron [2] that, after adequate supervised training, would be able to perform such classification.

However, according to the definition, the perceptron can perform the classification function only after supervised learning. Instead, Pavlov's dog must follow unsupervised learning. Currently, the mechanism, describing adequately unsupervised learning is Spike Timing Dependent Plasticity (STDP) [8–13]. It is to note, that the system, performing such kind of learning of Pavlov's dog, has been realized, using parylene-based memristive device. Details of this work will be presented in the successive section.

The first single-layer perceptron, based on memristive devices, was realized using titanium dioxide active layer and crossbar configuration of the electrodes [79]. Such configuration of the electrodes is widely used for different systems based on inorganic memristive devices [164–179].

The realized system has demonstrated the possibility to classify objects after adequate training. In particular, it was considered the possibility of using the system for the classification of graphic symbols.

Less than 1 year from the publication of the mentioned paper [79], a single-layer elementary perceptron was realized based on organic memristive devices – the main element of this book [180]. The main features of this system will be described in the adequate section.

Inorganic memristive devices have found interesting and unexpected applications in the field of biosensors [181–189]. In particular [185], a memristive device based on silicon semiconductors has been modified with sensitive biomolecules. Antibodies were immobilized at the surface of the nanostructure. Specific binding of antigens was shown to result in the shift of the position of the minimal current value with respect to the value of the applied voltage (X-axes of the voltage-current characteristics).

It has been shown [187] that the value of the shift depends on the antigen concentration in the solution.

The presented data confirm once more that voltage-current characteristics of memristive devices do not always demonstrate zero current at zero applied voltage. In the case of [185], the shift is due to the additional surface potential, appearing as a result of antigen-antibody interactions.

Finally, it is to note that memristive devices were fabricated also with silicon [67, 104, 190–193] – the main material of the current electronics technology.

1.5   Memristive Devices with the Organic Materials

The very first and most studied organic memristive device was the element based on the polyaniline layer in a contact with solid electrolyte on the basis of polyethylene oxide [47, 194]. We will not consider this device in this chapter, because it is the main element to which the book is dedicated. All aspects of its structure, fabrication methods, properties, and applications will be discussed in detail in appropriate sections. In this part, we will consider only other types of memristive devices using organic materials.

Probably, the most obvious use of organic materials for memristive devices is their utilization as plastic supports which allows to fabricate lightweight and flexible devices. In particular, in [195], it was realized a hybrid system containing eight pairs of "transistor-memristor," designed for use as flexible resistive-switching random

Fig. 1.7 Typical voltage-current characteristics of the realized memristive device [195] (arrows indicate the direction of the voltage variation). Inserts show the device structure and its equivalent circuit. (Reprinted with permission from: Kim et al. [195]. Copyright 2011 American Chemical Society)

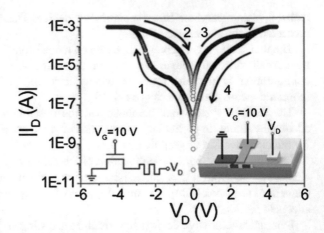

access memory. The scheme of each element of the system and appropriate voltage-current characteristics are shown in Fig. 1.7.

It is to underline that only one property of the organic materials was used in this work – flexibility.

Another obvious application of organic materials in memristive systems, including transistors, is the realization of transparent elements. In this case, the gate of transistors can be done from organic materials. In particular, P(VDF-TrFE) was used for these reasons in [196].

A less obvious and more interesting application of the organic materials for memristive devices is connected to the possibility of their use as an insulator capable to vary the resistance. A fibroin from Bombyx mori was used as such insulator in [197]. The insulator was placed between ITO and Al electrodes.

It was suggested [197] that the resistance switching mechanism in this structure is connected to the formation and destruction of conducting filaments in the fibroin layer. Scanning tunneling microscopy imaging in a constant height mode has been performed for the confirmation of this hypothesis. The switching of the device into a conducting state, in this case, is characterized by the formation of conducting filaments, while the back switching to the insulating state is characterized by the elimination of these filaments.

Formation and destruction of filaments were connected to the redox reactions in the fibroin layer resulting in the formation of SF^+ zones [197].

A completely organic system with properties of a memristive device was realized in [198]. Layers of poly (3,4-ethylene dioxythiophene): poly (styrene sulfonate) (PEDOT:PSS) were used as electrodes, while poly(4-vinylphenol) (PVP) used as an insulator, was placed between them. Resistance switching in this structure was studied in $-20 - +30$ V voltage range.

Measurements of the resistance were done by applying 15 V. This value of the applied voltage does not change the conductivity state of the system.

The realized system, however, has several serious disadvantages. First, the ON/OFF ratio of the system is only one order of magnitude. Second, it is necessary

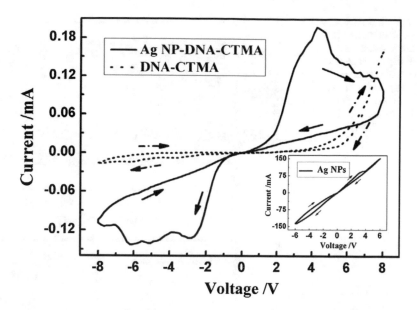

Fig. 1.8 Cyclic voltage-current characteristics of a memristive device where an active layer of DNA with the addition of cetyl trimethyl ammonium bromide and silver nanoparticles was placed between Al and ITO electrodes (for the comparison, the figure presents also characteristics, where the active layer was composed of pure DNA or silver nanoparticles) [199]. (Reprinted from Wang and Dong [199], Copyright (2014) with permission from Elsevier)

to apply rather high values of voltage, which limits seriously the areas of its applications. However, the system has one important advantage – it is completely organic.

Interesting memristive properties were observed in a system where a composite layer of DNA with the addition of cetyl trimethyl ammonium bromide and silver nanoparticles was placed between Al and ITO electrodes [199]. Typical cyclic voltage-current characteristics of this system are shown in Fig. 1.8 (for the comparison, the figure presents also the characteristic of the system done from pure DNA or silver nanoparticles).

Similarly, to the case of [198], also here it was necessary to apply rather high voltage for resistance switching. Moreover, these systems [199] revealed rather low stability and reproducibility that can be slightly increased using thermal treatment.

DNA was also used as an insulating layer with gold electrodes [200, 201]. The ON/OFF ratio in these cases was found to be more than one order of magnitude.

Interesting results were obtained when the egg albumin layer was placed between ITO and Al electrodes [202]. The ON/OFF ratio, in this case, was found to be more than three orders of magnitude. Moreover, the system revealed high stability both for the time of measurements and for the number of resistance switching cycles, which is shown in Fig. 1.9.

Fig. 1.9 Dependences of
the value of current in the
system in conducting and
insulating states on the
number of resistance
switching cycles (**a**) and
time of operation (**b**)
[202]. (Reproduced from:
Chen et al. [202])

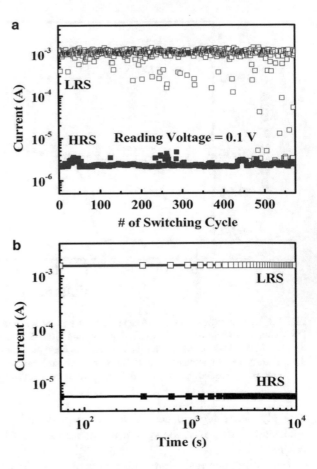

Resistance switching in these structures occurred after the application of the voltage about 2 V. Resistance switching mechanism was connected to the oxygen diffusion and electrochemical reactions, involving metal ions resulting in the formation and destruction of conducting filaments in the albumin layer.

Memristive behavior was also observed in Ti/PEDOT:PSS/Ti system [202]. The characteristics of such devices were found to be very unstable. However, the authors have done a rather unexpected conclusion – it can be useful for mimicking learning processes.

The high ON/OFF ratio (more than four orders of magnitude) was registered in a structure Pt/silver doped chitosan/Ag [204]. It is important that the required voltage values, in this case, were less than 1 V, as is shown in Fig. 1.10.

Despite the fact that the ON/OFF ratio of the structure is rather high, it has an important drawback connected to the instability and low reproducibility of the voltage value required for the resistance switching. It is more pronounced for switching from high to low conducting states. This behavior is connected to the mechanism of the resistance switching that is schematically shown in Fig. 1.11.

Fig. 1.10 Cyclic voltage-current characteristics of Pt/silver doped chitosan/Ag system (insert shows the device view) [204]. (Reprinted with permission from Hosseini and Lee [204]. Copyright 2015 American Chemical Society)

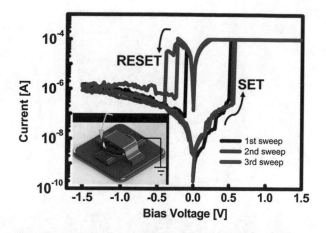

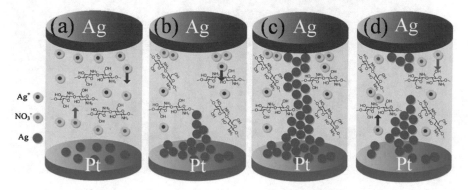

Fig. 1.11 Mechanism of the resistance switching in Pt/silver doped chitosan/Ag system [204]. (Reprinted with permission from Hosseini and Lee [204]. Copyright 2015 American Chemical Society)

As it is clear from Fig. 1.11, the mechanism is based on the formation and destruction of conducting silver filaments between electrodes. The random nature of the mechanism is responsible for the unreproducible values of the voltage required for the resistance switching [204].

Chitosan-containing layers were also used in systems with gold electrodes [205].

Among organic materials, parylene (poly-para-xylylene) attracts a high attention in the field of electronic devices including organic layers due to the simple and cheap production method of this polymer, its transparency, and the possibility of fabrication on flexible substrates [116, 206]. Moreover, parylene is an FDA-approved material and, therefore, it can be used for biomedical applications since it is completely safe for the human body; that is not the case for most polymers used in electronics [206–208]. In addition, parylene has found wide applications in electronics, including integrated circuits, thin-film transistors, lasers, waveguides, etc. [209, 210].

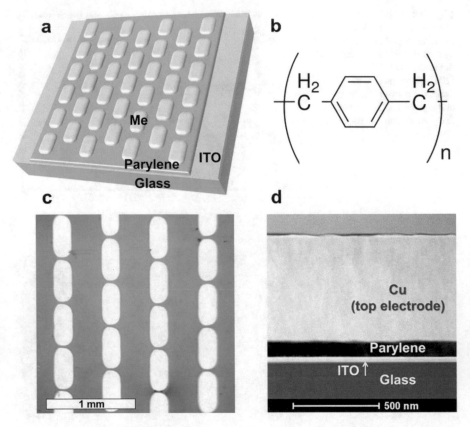

Fig. 1.12 (**a**) Scheme of the M/PPX/ITO memristive structure. (**b**) Parylene N repeat unit. (**c**) Microscopic image of Cu/PPX/ITO memristive elements, constituting the memristive structure (only a part of the sample is shown). (**d**) Layers of the Cu/PPX/ITO memristive element, a TEM image [211]. (Reproduced from: Minnekhanov et al. [211])

In [211], this material was used for the fabrication of memristive devices for their successive use for neuromorphic applications, in particular, for mimicking Pavlov's dog learning by the STDP algorithm. These results seem very important, therefore, we will consider them in more detail.

The scheme of the realized system is shown in Fig. 1.12.

Three types of parylene-containing systems with different electrodes were studied, namely, Ag/parylene/ITO, Cu/parylene/ITO, and Al/parylene/ITO. All of them revealed good memristive properties with ON/OFF ratio about 100, as is shown in Fig. 1.13.

Main features of all three studied systems are summarized in Table 1.1 [211].

Sample	max R_{off}/R_{on}	V_{set}, V	Number of resistance states	Endurance, cycles
Ag/parylene/ITO	100	2.1 ± 0.7	8	300
Cu/parylene/ITO	10,000	1.5 ± 0.5	16	1000
Al/parylene/ITO	100	2.7 ± 2.1	8	70

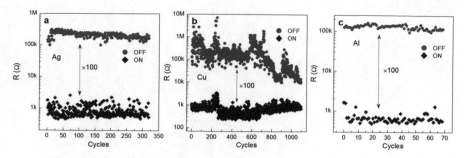

Fig. 1.13 Endurance of (**a**) Ag/parylene/ITO, (**b**) Cu/parylene/ITO and (**c**) Al/parylene/ITO memristors. Black and red points represent low- and high-resistance states, respectively. Pulse time was 100 ms; Uset $= -$Ureset $= 5$ V for (**a**), Uset $= -$Ureset $= 4$ V for (**b**) and Uset $= 5$ V, Ureset $= -8$ V for (**c**) [211]. (Reproduced from: Minnekhanov et al. [211])

Cu/parylene/ITO was chosen for the realization of the system for neuromorphic applications as it revealed the best characteristics, which is clear from Fig. 1.13 and Table 1.1.

The results related to the STDP-like learning of systems, based on these devices, as well as Pavlov's dog learning mimicking system [212–215], will be presented in Chap. 6.

In summary, it has been successfully demonstrated that parylene-based memristive devices are capable of learning (including learning using biologically inspired STDP-like rules) [211].

Chapter 2
Organic Memristive Device

This device was developed directly for mimicking important synapse properties in electronic circuits: capability to vary the conductivity (weight of connections), memory (short- and long-term), and rectification (unidirectional signal propagation). In this chapter, we will consider all important features of these devices: materials, architecture, working principles, and properties.

2.1 Basic Materials

The working principle of the organic memristive device is based on the property of polyaniline to vary significantly its conductivity according to the form in which it exists [216].

Polyaniline can be in different forms starting from completely reduced (pernigraniline) to completely oxidized (leucoemeraldine) ones as is shown in Fig. 2.1.

For practical applications, the most important form is an emeraldine one, as only it allows to have high conductivity values. For this reason, the sample, containing polyaniline, must be doped, which transfers insulating emeraldine base form into a conducting emeraldine salt one.

Considering the working principle of the device under the consideration, it is very important to have the possibility to vary the conductivity of an active layer not only by doping-dedoping processes but also by electrochemical redox reactions. Polyaniline provides this possibility.

Processes, responsible for the conductivity variation, occurring in polyaniline are schematically shown in Fig. 2.2.

As was mentioned above, the emeraldine base (EM in Fig. 2.2) form is a starting material for the device fabrication. This form is an insulating one. It is necessary to perform a doping process for its transfer into conducting emeraldine salt form. Usually, doping is done by the acid treatment (in the simplest case, it is an immersion

© Springer Nature Switzerland AG 2022
V. Erokhin, *Fundamentals of Organic Neuromorphic Systems*,
https://doi.org/10.1007/978-3-030-79492-7_2

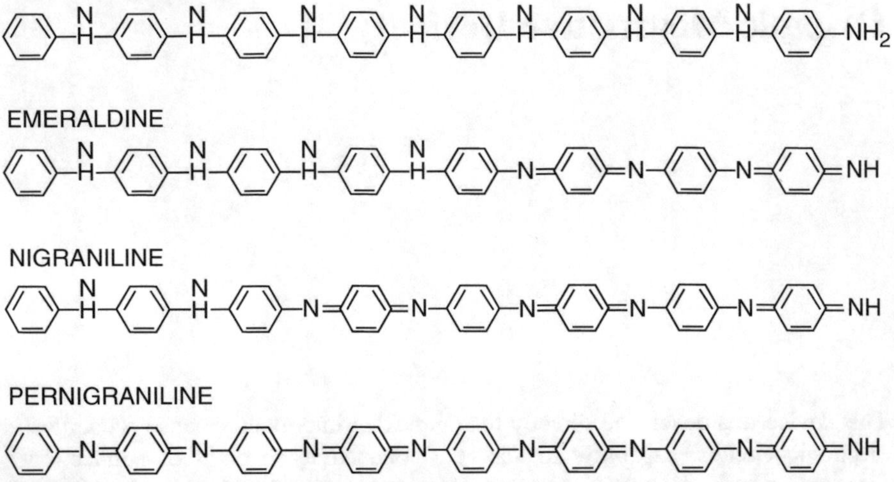

LEUCOEMERALDINE

EMERALDINE

NIGRANILINE

PERNIGRANILINE

Fig. 2.1 Different forms of polyaniline [216]. (Reprinted from Kang et al. [216], Copyright (1998) with permission of Elsevier)

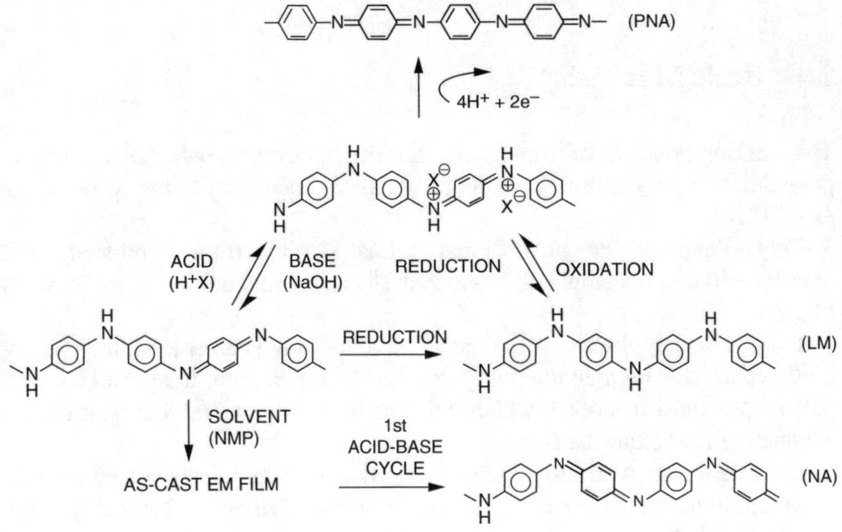

Fig. 2.2 Reactions responsible for the conductivity variations of polyaniline [216]. (Reprinted from Kang et al. [216], Copyright (1998) with permission of Elsevier)

into hydrochloric acid solution or treatment with its vapors). As a result, the polymeric chain becomes protonated, and a non-compensated charge (hole) appears in it. For providing the electrical neutrality of the whole molecule, the oppositely charged acidic ions come closer to the chain (in the simplest case, it is Cl$^-$ ion).

As is clear from Fig. 2.2 (left part), the doping process is a reversible one – treatment with basic solutions will result in the transferring of the polymer into an insulating form. Practically, dedoping occurs even if the sample is immersed into an aqueous solution and even in air (which is likely due to the presence of water vapors). This is, of course, a negative phenomenon that results in the variation of the polyaniline conductivity in time. This problem was resolved by using special doping agents and the application of protecting covering layers, which will be discussed in the appropriate section.

The working principle of the device is based on electrochemical redox reactions, shown in the right part of Fig. 2.2. These reactions are reversible ones, and the conductivity ratio of polyaniline in oxidized and reduced states is about eight orders of magnitude. Therefore, applying oxidizing or reduction potentials, it is possible to change reversibly the conductivity state of polyaniline between insulating and conducting ones. Oxidizing potential of polyaniline in bulk is about +0.3 V, while the reduction one is about +0.1 V [217].

During redox reactions, it is not enough to provide or remove electrons from the chain; it is necessary to compensate also the charge of the chain by electrostatic interactions with ions of opposite charge sign. Therefore, for the construction of the device, based on the variation of polyaniline conductivity due to redox reactions, it is necessary to have a medium where these reactions can occur. Therefore, the device must contain a zone of electrolyte in a contact with polyaniline layers. Ideally, the electrolyte must be in a solid form for more effective use of these structures for electronic device production.

Polyethylene oxide was chosen as a matrix for solid electrolyte fabrication. Its formula is presented below:

$$(-CH_2CH_2O-)_n$$

The choice of this material was determined by the fact that it was very successfully used for the production of lithium batteries [218] and highly effective capacitors [219].

Salts of lithium were used as a source of ions in the device, as lithium, due to its small sizes, has the highest mobility especially in solid media. In particular, we have used lithium perchlorate, because lithium chloride is a very hygroscopic material, and polyethylene oxide layers, containing LiCl, are always in liquid or gel form, adsorbing water vapors from the atmosphere.

The model, describing ionic transport in this material, is based on the jump transition of ions between allowed states, connected to the oxygen atoms, coordinating lithium ions.

2.2   Structure and Working Principle of the Device

As was mentioned above, the main mechanism responsible for the device function is connected to redox reaction in the active layer of polyaniline. Therefore, the device must contain layers of polyaniline with electrodes, connected to them, between which the value of the conductivity must be controlled (the value of the current measured at a certain fixed value of the applied voltage). The thickness of the active layer is a very critical parameter. On the one hand, it must allow rather high conductivity of the device for the reliability of measurements (therefore, it must be rather thick). On the other hand, the layer must be rather thin, as the device functioning implies the diffusion of ions in the solid-state phase. Therefore, the transfer of the whole material of the active layer from one conductivity state to the other within a rational period of time can be effectively done only if the thickness of this layer is rather small.

For this reason, all initial devices and a significant part currently used were fabricated using the modified Langmuir-Schaefer technique [220] for the deposition of active polyaniline active layer. The method is based on the formation of a monomolecular layer at the air/water interface with their successive transfer onto solid supports. The state of the monolayer, controlled by measuring its surface pressure, can be varied by the compression with barriers. The method is well known since the beginning of the twentieth century [221–223]. Thus, we will not consider it in detail, but indicate only the main steps and their differences with respect to the "classic" method.

The original version of the method implied the spreading of the substance solution at the air/water interface in such a way that molecules in the monolayer are in a very expanded ("two-dimensional gas") state. The monolayer is compressed with barriers to the condensed state and then transferred onto solid supports. The most widely used technique of the monolayer transfer is called the Langmuir-Blodgett method, when the solid support is moving vertically through the compressed monolayer.

Modification of the method called the Langmuir-Schaefer technique, when the compressed monolayer was successively touched by a solid support, was used for the fabrication of the active layers of organic memristive devices. The method allows to decrease significantly the deposition time, and it was successfully used for different applications, in particular, for the deposition of a layer containing protein molecules [224–229].

An important difference with respect to the "classic" version of the Langmuir-Schaefer technique is that after the compression of the monolayer and reaching the target surface pressure, the barriers are stopped and a grid with the cells sizes, corresponding to those of the sample, is placed on the monolayer. The sample is successively touching the monolayer in each cell with the removal of the water drop with an air jet after each monolayer transfer. This separation with the grid is necessary for providing homogeneous covering of the sample surface with deposited film. In fact, without this separation, each touching would result in the fact that not

Fig. 2.3 Symbol and schematic representation of the organic memristive device

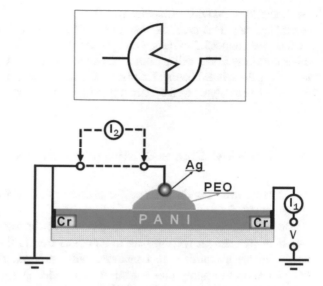

only the monolayer in direct contact with the sample but also that on the top of the water drop with not controllable structure will be transferred, which would result in inhomogeneity of the deposited layer.

Active polyaniline layers of fabricated devices contained usually 12–60 transferred monolayers. The thickness of one monolayer, measured by different methods, was found to be about 1 nm.

Schematic representation of the organic memristive device and its symbol are shown in Fig. 2.3. The symbol is different from that used for memristors in literature. There are two reasons for it: first, it represents better anisotropy of the conductivity; second, it was introduced before the HP work of 2008 [48], when the word "memristor" was practically unknown for researchers.

A thin layer of polyaniline connects two metal electrodes, previously evaporated on an insulating support. In the central part, the polyaniline layer is covered by a stripe of solid electrolyte (lithium salt doped polyethylene oxide), forming an active zone, where possible electrochemical redox reactions can occur. The concentration of the lithium ions in the electrolyte is a critical parameter. On the one hand, this concentration must be rather high for allowing effective redox reactions. On the other hand, it must be as small as possible, because the ionic current, being a part of the total current of the device, will diminish such important characteristics as the ON/OFF ratio of the device characteristic in conducting and insulating states.

As redox reactions begin to go on only at certain values of the potential, we must have a reference point, with respect to which the potential of the active zone will be maintained and/or varied. For this reason, a thin metal wire should be connected to the solid electrolyte stripe. This wire is referred to as reference (or, as it was called in early works, gate) electrode. Different materials were tested for the fabrication of the reference electrode. In particular, gold, silver, and platinum were tested [47]. The best results have been obtained when silver was used for the fabrication of this electrode. In the device, this wire was directly connected to one of the two metal electrodes, called "source" in analogy with field-effect transistors. Both these electrodes were connected to the ground potential, while the voltage was applied to the other metal electrode, called "drain." In such a way, the applied potential is distributed along the whole channel length and the active zone of the contact of polyaniline with solid electrolyte has a certain potential with respect to the reference electrode.

2.3 Electrical Characteristics of the Device

The scheme used for the electrical characterization of the memristive devices is shown in Fig. 2.4.

Despite the fact that the organic memristive device can be considered as a two-terminal element with respect to its connection to the external circuit (source and reference electrodes are connected with each other), the scheme shown in Fig. 2.4 allows to apply one voltage to the device and to measure two values of currents: the total current of the device and the current in the circuit of reference electrode that can be directly connected to the ionic current in solid electrolytes. The difference between these measured currents corresponds to the electronic current, passing through the active layer of polyaniline.

The first step of the device characterization was measurements of their cyclic voltage-current characteristics. The measurements were done in the following way. Initially, the device was maintained at zero applied potential for about 5 minutes. After it, successively increasing voltage steps were applied, usually with an increment of 0.1 V. For each value of the applied voltage, a time delay was used (from several seconds to several minutes) for coming to the equilibrium state, and the readout of the current value was done after this delay. The maximum value of the

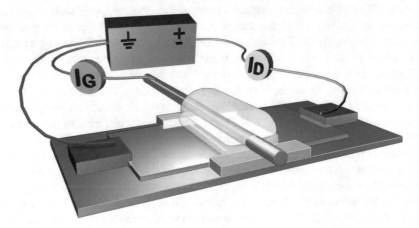

Fig. 2.4 The scheme used for the electrical characterization of organic memristive devices. (Reprinted from Berzina et al. [249], with the permission of AIP Publishing)

applied voltage did not exceed 1.5 V for avoiding overoxidation processes transferring irreversibly polyaniline into an insulating state. After reaching the maximum potential, the applied voltage was decreased with the same step and delay till reaching the maximum negative potential value. After it, the applied voltage was returned to zero value with the same step and delay.

Typical voltage-current characteristics are shown in Fig. 2.5.

Let us consider in more detail the characteristics shown in Fig. 2.5. For better understanding, we will concentrate the attention not on the dependence of the total current, but on the ionic and differential (electronic) ones. The first one is responsible for the ionic drift between layers of the conducting polymer and solid electrolyte, providing the redox reactions, while the second one determines the conductivity of the polyaniline active layer and the whole device.

The measurements were started at zero potential, and its increment was performed in the positive direction. At low values of the applied voltage, we can see the low conductivity of the device (both for the total current and its electronic component). This behavior is due to the fact that polyaniline is in its reduced insulating form at these values of the applied voltage. The situation remains the same till the point when the applied voltage reaches the value of about +0.5 V. After it, we can see a significant increase in the conductivity for the electronic current that is accompanied by the appearance of a positive peak in the characteristics for the ionic current. Polyaniline in the active zone starts to transfer itself into an oxidized conducting state. The process of the transformation to the oxidized state requires time. Thus, if the voltage variation rate is rather fast, the complete oxidation of the whole polyaniline layer can be not reached even when the maximum value of the voltage was applied. Further increase of the voltage is characterized by the high conductivity of the device that remains also during the successive decrease of the applied voltage. Switching of the device to the insulating state occurs when the value of the applied voltage begins to be less than +0.1 V, which corresponds to the reduction of polyaniline in the active zone. The device remains in the insulating state for all values of the applied negative voltages.

The values of the applied voltages, when the resistance switching of the device occurs, are different from the oxidizing and reduction potentials of polyaniline in the bulk, which is connected to the fact that the voltage is applied not directly to the active zone but to the drain electrode, distant from this zone. The actual value of the active zone potential is different from the applied voltage as the last one is distributed along the whole channel length. This behavior will be considered in more detail in the section, dedicated to the description of the device operation model.

An important feature of each memristive device is a variation of the hysteresis loop shape for different delay times at each applied voltage value (voltage scanning frequency). Similar tendency has been observed also for our device. Dependences of the electronic and ionic currents on the applied voltage for different time delays are shown in Figs. 2.6, 2.7, and 2.8.

When the delay time is rather small (high voltage scanning frequency), polyaniline in the active zone has not enough time to be transferred into the conducting state even when the applied voltage reaches its maximum value. This

Fig. 2.5 Cyclic voltage-current characteristics of organic memristive devices, measured with the voltage step of 0.1 V and time delay of 60 seconds: (**a**) total current in the drain chain, (**b**) ionic current in the reference electrode chain, (**c**) differential current (electronic current in the polyaniline channel). Empty rhombs correspond to the voltage-increasing branches (absolute values); filled squares correspond to the voltage decreasing branches (absolute values). (Reprinted from Erokhin et al. [47], with the permission of AIP Publishing)

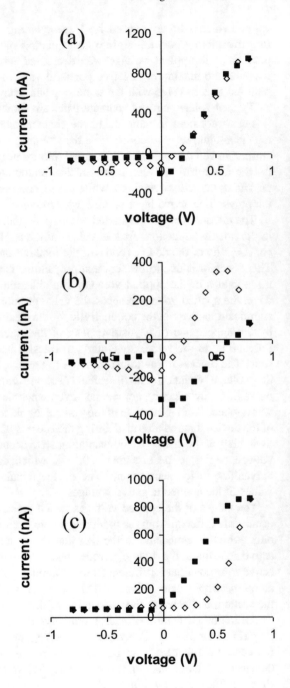

is the reason that the hysteresis loop for positive values of the applied voltages is absolutely not pronounced (more pronounced hysteresis in the negative branch of the voltage scan results from the fact that accumulated effect of the application of the

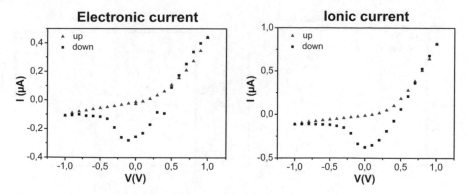

Fig. 2.6 Cyclic voltage-current characteristics of organic memristive devices for electronic and ionic currents. Time delay is 5 s. (Reprinted from Dimonte et al. [339], Copyright (2014), with permission from Elsevier)

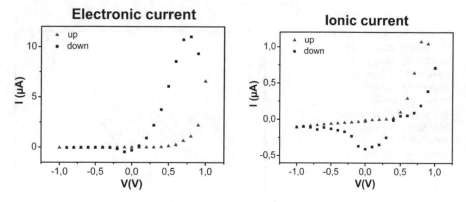

Fig. 2.7 Cyclic voltage-current characteristics of organic memristive devices for electronic and ionic currents. Time delay is 20 s. (Reprinted from Dimonte et al. [339], Copyright (2014), with permission from Elsevier)

potential, lower than the reduction one, is much more than in the case of oxidation due to the application of the appropriate positive potential) (Fig. 2.6).

In the case of the increased time delay, it is possible to observe an interesting feature in the cyclic characteristic for the electronic current (Fig. 2.7): in the back scan of the positive branch of the voltage variation, we can see a negative differential resistance (the value of the current increases while the applied voltage decreases in the range from +1.2 V till +0.7 V). Such behavior is connected to the fact that the time delay, used in this experiment, is not enough for transferring the whole polyaniline in the active zone into the conducting oxidized state. During the back scan of voltage and before reaching the reduction potential, the polymeric material still continues its transfer into the conducting state. Initially, in the range +1.2 V – +0.7 V, the increase of the channel conductivity is more pronounced than

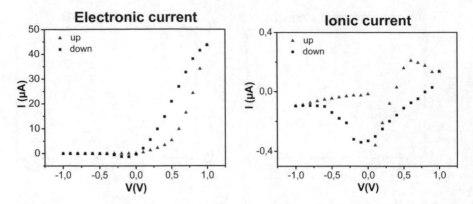

Fig. 2.8 Cyclic voltage-current characteristics of organic memristive devices for electronic and ionic currents. Time delay is 60 s. (Reprinted from Dimonte et al. [339], Copyright (2014), with permission from Elsevier)

Fig. 2.9 The dependence of the ON/OFF ratio of the conductivity of organic memristive devices on the value of the delay time of the voltage application at each step. (Reprinted from Dimonte et al. [339], Copyright (2014), with permission from Elsevier)

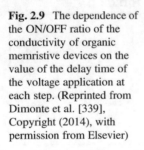

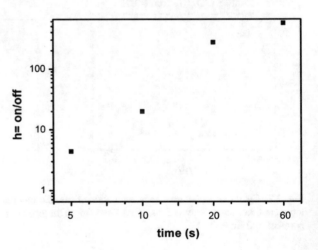

the Ohmic decrease of the current (supposing that the resistance does not vary). Therefore, till about +0.7 V, we see the increase of the current, while the voltage is decreasing, which is due to the fact that the process of the polyaniline transfer into the conducting state continues even during the decrease of the applied voltage value.

When the time delay is 60 s, all oxidizing processes are finished before the applied voltage reaches its maximum value and cyclic voltage-current characteristics obtain their "classic" form (Fig. 2.8).

Variation of the delay time changes not only the shape of the hysteresis loop but also the ON/OFF conductivity ratio in conducting and insulating states of the device, as is shown in Fig. 2.9.

Characteristics for the electronic (differential) current, presented in Figs. 2.6, 2.7, and 2.8, have two important features: hysteresis and rectification. Both of them are

Fig. 2.10 Temporal
dependences of the total
current of the organic
memristive device, when the
voltage values of +0.6 V (**a**)
and −0.1 V (**b**) are applied
to the drain electrode.
(Reprinted from Erokhin
et al. [47], with the
permission of AIP
Publishing)

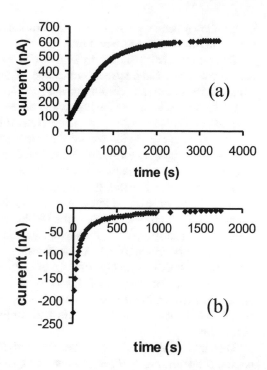

very important for the realization of electronic devices mimicking synapse properties. However, for a better illustration of this statement, it seems useful to consider transition kinetics (variation of the current) of the conductivity variation at fixed values of the applied voltage.

Temporal dependences of the variation of the total current in the drain circuit of the organic memristive device when the voltage values of +0.6 V and −0.1 V are applied are shown in Fig. 2.10.

Let us first consider the characteristics shown in Fig. 2.10a. When the positive voltage with a value higher than a certain value, necessary for the oxidation of polyaniline in the active zone, is applied (in our case, it is about +0.5 V), a gradual increase of the device conductivity takes place. Such behavior is similar to that of synapses, described by the Hebbian rule in its "classic" formulation [7], that can be also connected to the term "unsupervised learning." This rule, cited in the introduction section, can be reformulated for the electronic circuits in the following way: electronic circuit elements (conductors and contacts) must increase their conductivity with the duration and/or frequency of their involvement in the formation of the signal transferring pathways. In fact, let us imagine a system containing inputs and outputs and having a complicated architecture of the medium between them, composed of devices, with properties, described above. The learning of this system will mean the induction of short- and long-term connections between certain inputs and outputs (or groups of inputs with outputs). As more frequently the formed chains of the signal transfer will be used, as more likely it will be used again for the information processing when the configuration of stimuli at input electrodes will be similar to the already processed one.

The characteristics shown in Fig. 2.10b are also very important. First, if we consider a complicated system composed of organic memristive devices with the conductivity variation according to data, presented in Fig. 2.10a, earlier or later, it will come to a saturation situation when all elements will be in a high conducting state and no further learning and adaptation will be possible. Therefore, the dependence, shown in Fig. 2.10b, can be used for bringing the system the state out of the equilibrium. Periodic short-term application of negative potentials between all input-output pairs will decrease the conductivity level of all elements of the system below the saturation level. Second, the characteristics shown in Fig. 2.10b can establish a basis for the so-called supervised learning. In fact, a complicated system, due to the unsupervised learning according to the Hebbian rule, can establish some connections between input and output electrode pairs, establishing some classification algorithms that are a priori wrong. In order to retrain the system, it is necessary to eliminate the formed chains of the signal transfer. In the case when the system is composed of the organic memristive devices of this type, it will be enough to invert the applied potentials between input-output pairs, having the wrong association. The induced signal transfer chain will be inhibited according to the characteristics, shown in Fig. 2.10b.

The significant difference in the conductivity switching time constants of processes, shown in Fig. 2.10, is an interesting and, from the first glance, not obvious phenomenon. Quantitative consideration of such behavior will be presented in the section, dedicated to the model of the organic memristive device function. Here, we restrict ourselves only to the qualitative explanation of this behavior.

In the case shown in Fig. 2.10b, a negative voltage (any) is applied to the device in the conducting state. As the device is in a conducting state, we can see the homogeneous distribution of the potential along the channel length. Therefore, the whole active zone will be at a negative potential with respect to the reference electrode that is maintained under the ground potential. Taking into account that the reduction potential is about +0.1 V, we can claim that the reduction of polyaniline and its transfer into the insulating form will occur simultaneously in the whole active zone.

The situation is absolutely different when we transfer the device from its insulating form to a conducting one. Figure 2.11 illustrates the processes responsible for this transfer.

As the initial state of the device in this case is an insulating one, the voltage, applied to the drain electrode, will be distributed mainly along the part of the polyaniline channel in the active zone (parts of polyaniline layer, not in contact with solid electrolyte, are in the oxidized form, and therefore, their conductivity is at least two orders of magnitude higher than that in the active zone). Let us suppose that we have applied +0.6 V. Considering the homogeneity of the active zone, we can suppose that the potential is linearly distributed on its length, as it is schematically shown in Fig. 2.11a. Therefore, only half of the polyaniline layer in the active zone (closer to the drain electrode) will have a potential higher than the oxidizing one

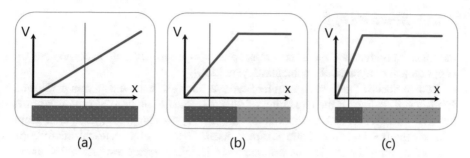

Fig. 2.11 Temporal evolution of the potential distribution profile along the active zone of the organic memristive device after the application of +0.6 V to the device in the insulating state

(+0.3 V). Thus, the transfer of polyaniline into the conducting state will occur only in this area. After this transfer, the applied voltage will be redistributed, as it is schematically shown in Fig. 2.11b. In this case, half of the remaining reduced polyaniline layer in the active zone will be at a potential higher than the oxidizing one, and therefore, it will be transferred into a conducting state. The process will be continued till the whole polyaniline in the active zone will be transferred into a conducting state. Obviously, the process is continuous and not discrete. Moreover, as the device has rather large sizes of the active zone (20 microns in the best case), the velocity of the ion motion in the electrolyte (efficiency of the resistance switching) is a function of time and linear coordinate (from source to drain) of the part of polyaniline channel.

Summarizing the above considerations, it is possible to conclude that the difference in the kinetics of the device switching from its conducting state to an insulating one and back is connected to the fact that in the first case, we have the simultaneous transfer of the whole polyaniline layer in the active zone to its reduces form, while in the case of the back transition, we have a gradual displacement of the boundary of the conducting area in time in the direction from drain to source.

2.4 Device Working Mechanism

Two sets of experiments were carried out for having the possibility of suggesting the device's working mechanism. These experiments were done using optical, Fourier transform infrared, and Raman spectroscopic methods and X-ray fluorescence. Let us consider these methods and the results obtained by their applications.

2.4.1   Spectroscopy

Samples for optical and FT IR spectroscopies were deposited onto grass and silicon supports and contained 48 molecular layers [230].

Raman spectra have been acquired with a micro-Raman spectrometer (Horiba Jobin Yvon) using for the excitation of different lines of an argon-krypton laser. It allowed to register and confirm resonant effects, varying the wavelength from 456 nm (argon line) to 647 nm (krypton line). The use of a confocal microscope allowed to study spectra in microscopic areas (minimal space resolution was about 1 micron), to reconstruct spectral maps along the working channel, and to study the dependence of the spectra on the thickness of the heterojunction (using height scans). One critical problem was faced during performing these experiments: degradation of the sample under the illumination, resulting in the change of the spectra during successive measurements. In order to avoid this effect, a special study has been done for the estimation of the max power of the incident beam that does not induce the sample damage. Usually, the power did not exceed 0.3 mW. It resulted, of course, in a significant increase in the acquisition time but provided a good reproducibility of the experimental results.

Initial variation of the polyaniline conductivity occurs during doping (transfer from emeraldine base to emeraldine salt form). It is accompanied not only by the significant increase of the conductivity but also by the variation of optical absorption spectra due to the variation of the relative amount of quinon-benzen groups in the polymer chain.

Optical absorption spectra of conducting and insulating forms of polyaniline are shown in Fig. 2.12.

The presence of a high level of background in the near UV region is due to the polaron absorption [231, 232] that is also visible in Fourier transform infrared spectra (FT IR), shown in Fig. 2.13.

To complete the results, FT IR spectra of doped and undoped polyethylene oxide are shown in Fig. 2.14.

Before considering Raman and IR spectra, let us discuss the optical ones. As it is clear from Fig. 2.12, polyaniline spectra for conducting and insulating states are very different. This property has created a basis for a contactless technique of the estimation of the polyaniline conductivity in different zones [233], which, as it will be shown in appropriate sections, allows to understand better the function of circuits and systems, based on organic memristive devices.

These measurements were done directly on the working device. For simplicity, the measurements of the reflection spectra, instead of the absorption ones, have been performed, using the configuration, schematically shown in Fig. 2.15.

The layer of polyaniline was deposited onto glass transparent support between two metal electrodes (source and drain). The structure was placed into contact with polyethylene oxide-based electrolyte in a gel state, placed in the well, formed in the Teflon bulk substrate, containing also the reference electrode (silver wire). The channel of polyaniline was partially covered by Kapton or photoresist in order to

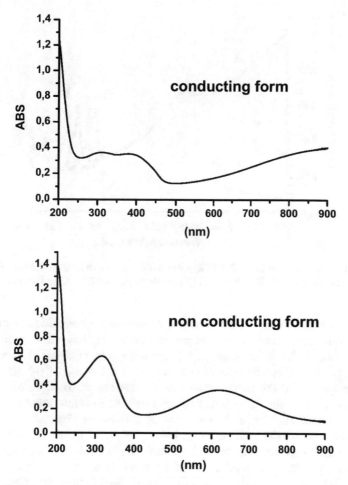

Fig. 2.12 Optical absorption spectra of conducting (upper panel) and insulating (lower panel) forms of polyaniline [230]. (Reprinted from Berzina et al. [230], with the permission of AIP Publishing)

localize the active zone and isolate the rest of the channel from the direct contact with electrolyte. Particularly, the channel length between electrodes was 15 mm, and the active zone area was 100 mm². Reflection spectra of three forms of polyaniline in the device are shown in Fig. 2.16.

The reflection spectrum of the initial state of the polyaniline channel has been acquired when the electrical circuit was disconnected and the electrolyte pH was equal to 7.0. Spectra corresponding to the conducting and insulating forms of polyaniline in the active zone (pH 7.0 of the electrolyte) were acquired when +0.6 V and −0.2 V were applied to the drain electrode, respectively. Spectra presented in Fig. 2.16 demonstrate that registration of the optical spectra can show the state of polyaniline in the active zone: doped-dedoped, oxidized-reduced.

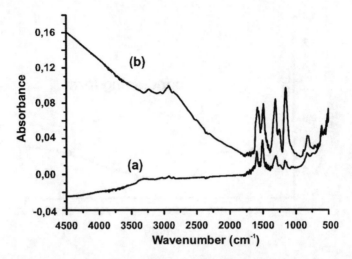

Fig. 2.13 Fourier transform infrared (FT IR) spectra of insulating (**a**) and conducting (**b**) forms of polyaniline. (Reprinted from Berzina et al. [230], with the permission of AIP Publishing)

As was mentioned above, the presented results on the correlation of the optical and electronic properties are very important. However, important results for the understanding of the device function principles have been obtained using Raman spectroscopy [230]. A characteristic feature of the method is the fact that the scattering spectra for the same sample can be different if the light of a different wavelength is used for their excitation. In order to illustrate this statement, the spectra of conducting and insulating forms of polyaniline obtained using 488 nm and 647 nm lines for the excitation are shown in Figs. 2.17 and 2.18, respectively.

The presented spectra indicate the combined effects of the doping and resonance conditions. Therefore, these effects must be taken into account during the result interpretation. However, as it is clear from the presented figures, the effect of doping is more pronounced in the range of 1500–1600 cm^{-1} when the excitation is done with a light at a 488 nm wavelength. A strong peak at 1508 cm^{-1} for the insulating state is practically absent for the conducting form. It is to note that a similar difference was observed also in the case of IR spectra, which allows to exclude the influence of other resonant phenomena that can make their contribution in the variation of spectra shape for different forms of polyaniline. This behavior was explained by the variation of the quinone/benzene ratio in the chain of the polymer chain [234].

Raman spectra of the just prepared device (before the final doping) in different points before the application of any potential to the drain electrode are shown in Fig. 2.19.

The spectra are significantly different in the range 150–1600 cm^{-1}, and a peak of perchlorate is explicitly visible at 930 cm^{-1}. This is the first indication of the presence of the Li$^+$ motion through the interface polyaniline-polyethylene oxide.

Raman spectra of the device after the second final doping are shown in Fig. 2.20.

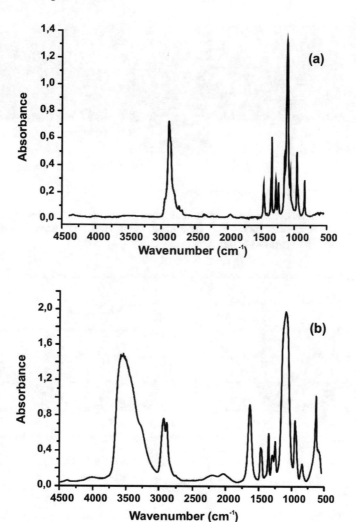

Fig. 2.14 Fourier transform infrared (FT IR) spectra of pure (**a**) and LiClO$_4$ doped (**b**) polyethylene oxide. (Reprinted from Berzina et al. [230], with the permission of AIP Publishing)

The spectra shown in Fig. 2.20 were acquired near the drain electrode (a), in the active zone of polyaniline-polyethylene oxide junction, but far from the silver electrode (b), and near the silver electrode (c). Comparison of this figure with Fig. 2.19 allows to conclude that the reduction of polyaniline in a contact with polyethylene oxide still occurs, but it is not as pronounced as it was before the secondary doping. It is also to note that the peak (930 cm^{-1}) corresponding to the perchlorate is absolutely absent in the area, where there is no polyethylene oxide and its intensity is maximal near the silver wire. It is important that the intensity of this peak in the area closed to the silver wire is at least 5 times higher with respect to all

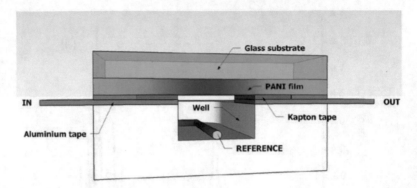

Fig. 2.15 The scheme of the experiment for registering the variations of the reflection spectra of organic memristive devices during their functioning [233]. (Reprinted from Springer Nature, Pincella et al. [233], Copyright (2011))

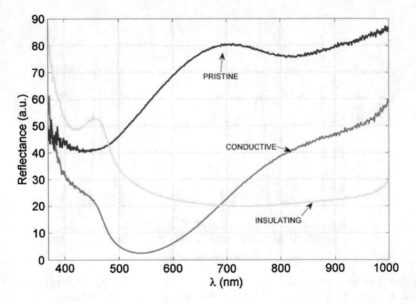

Fig. 2.16 Reflection spectra of polyaniline active layer: electrolyte at pH 7.0 (emeraldine base); conducting and insulating form of polyaniline in the active zone (emeraldine salt, electrolyte at pH 4.0), when voltages of +0.6 V and −0.2 V were applied to the drain, respectively [233]. (Reprinted from Springer Nature, Pincella et al. [233], Copyright (2011))

other places in the active zone. This phenomenon is likely connected to the effect of surface-enhanced Raman scattering (SERS) [235–237].

Let us consider now the variation of the device Raman spectra when two characteristic voltage values, responsible for the resistance switching, are applied, namely, +0.6 V and −0.2 V.

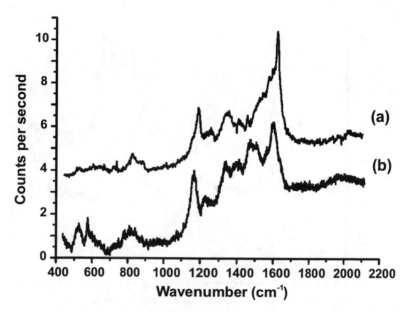

Fig. 2.17 Raman spectra of the conducting form of polyaniline obtained using optical lines of 488 nm (**a**) and 657 nm (**b**) for the excitation [230]. (Reprinted from Berzina et al. [230], with the permission of AIP Publishing)

Raman spectra, measured in different time moments in a point near the silver electrode in the wavelength range, corresponding to the position of the perchlorate peak, are shown in Fig. 2.21.

The spectra shown in Fig. 2.21 correspond to the completed cycle of the organic memristive device function, starting from the pristine form, measured immediately after the application of −0.2 V and 15 and 30 minutes after it. After it, the applied voltage was switched to +0.6 V, and the temporal evolution of the system spectra is illustrated by successive three curves (Series II). After it, the voltage was switched again to −0.2 V (Series III). The cycles were repeated several times (Series IV and successive (not shown)). Even if the tendencies were similar, some differences were observed. It is to note that the transition time was rather long and the transfer from the conducting state to the insulating one was faster than the back transition, in complete agreement with previously reported data on the electrical properties.

It is interesting to consider the behavior of the perchlorate peak. Its shape changes from an asymmetric single maximum to a double peak when the sign of the applied voltage is changed. A single peak corresponds to the presence of perchlorate ions in the solid electrolyte, while the double peak corresponds to the crystalline form of the $LiClO_4$ [230]. Thus, the spectra shown in Fig. 2.22 indicate the formation and destruction of $LiClO_4$ clusters during cyclic application of the voltages of the opposite sign. These transitions were never observed if no voltage was applied – the peak at 930 cm^{-1} was never transferred into the double peak form.

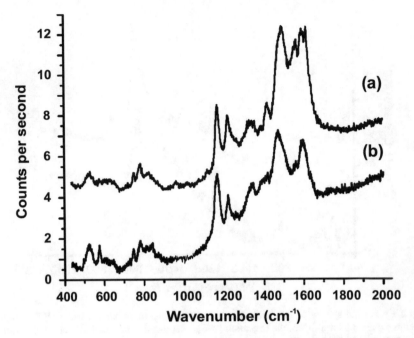

Fig. 2.18 Raman spectra of the insulating form of polyaniline obtained using optical lines of 488 nm (**a**) and 657 nm (**b**) for the excitation [230]. (Reprinted from Berzina et al. [230], with the permission of AIP Publishing)

The presented data allowed to suppose that the resistance switching is due to the redox reactions in the active zone of the contact of polyaniline layer with polyethylene oxide-based electrolyte. These reactions are accompanied by the motion of Li^+ ions between polyaniline and polyethylene oxide.

When the potential of -0.2 V is applied to the drain electrode, Li^+ ions penetrate the polyaniline layer and participate in the reduction process, blocking its conductivity. As the reduction potential has a positive value, this penetration occurs even when no voltage is applied to the device. Due to the difference in sizes and, therefore, in the mobility of Li^+ and ClO_4^- ions in the solid phase, we can suppose that only Li^+ ions are effectively involved in processes responsible for the conductivity variation. Application of the negative potential results in the displacement of lithium ions to the polyaniline. Therefore, lithium perchlorate complex cannot be formed, and we observe an asymmetric singlet peak of perchlorate ions in the Raman spectrum. The application of $+0.6$ V results in the fact that, after some time, Li + ions enter back to the polyethylene oxide-based electrolyte. The increased lithium concentration results in the formation of lithium perchlorate microcrystals, and the corresponding Raman peak is transferred into splitting of the previously observed maximum, as is shown in Fig. 2.21.

Summarizing the results of this section, it is possible to conclude that spectroscopic methods allow to monitor the conductivity state of the organic memristive

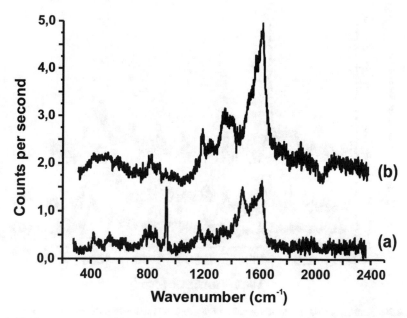

Fig. 2.19 Raman spectra of polyaniline-polyethylene oxide heterojunction before the final doping with HCl. Spectra were measured near the reference electrode (silver wire) (**a**) and in the area where only polyaniline was present (**b**) [230]. (Reprinted from Berzina et al. [230], with the permission of AIP Publishing)

device in the active zone of the contact of polyaniline and solid electrolyte, induced by redox reactions, accompanied by the lithium ions diffusion. These results established a basis for the model describing experimentally measured characteristics of such devices. In particular, it was for the first time explained why the deposition of solid electrolyte layer, based on polyethylene oxide, results immediately in the significant decrease of the polyaniline conductivity in the active zone: lithium ions penetrate polyaniline layers due to the concentration gradient and participate in the reduction reaction, as the reduction potential of polyaniline has a small positive value. It was also shown that the application of the voltage between the source and drain electrodes results not only in the appearance of the current in the channel but also in the current in solid electrolytes, whose temporal integral (transferred ionic charge) is responsible for the actual conductivity state of the entire device. Finally, the presented data have confirmed at the structural level the results on the different transfer kinetics of the device switching from the conducting state to an insulating one and back. More direct experiment on the ion motion during the device operation will be considered in the successive section.

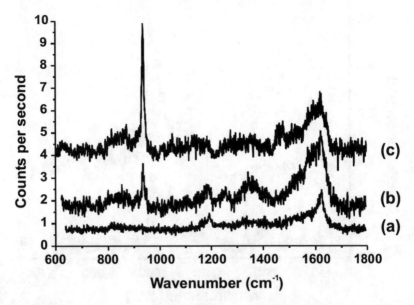

Fig. 2.20 Raman spectra of the active zone of the organic memristive device, measured after the secondary doping: (**a**) in the area, closed to the drain electrode; (**b**) in the active zone of the polyaniline-polyethylene oxide contact, but far from the silver electrode; (**c**) in the active zone of the polyaniline-polyethylene oxide contact near the silver electrode [230]. (Reprinted from Berzina et al. [230], with the permission of AIP Publishing)

2.4.2   X-Ray Fluorescence

The results presented in the previous section allowed to suggest that the motion of ions in the active zone between polyaniline and electrolyte is very important for the resistance switching of the organic memristive device.

The experiment that we will consider now was planned and carried out for the direct confirmation of this hypothesis.

X-ray fluorescence technique was chosen as a method, allowing to register a presence of certain ions in the localized space [238, 239]. However, the possibility of the application of this technique has demanded to vary experimental conditions. First, it was necessary to use synchrotron radiation for having the high-energy lines and adequate intensity for making real-time experiments. Therefore, the experiment was carried out at the ID10 beamline of European Synchrotron Radiation Facilities (ESRF, Grenoble). Second, the variation of the ions concentration during the device function should be localized in a thin layer of the active channel. Third, the fluorescence of lithium cannot be practically registered in normal conditions. Therefore, it was substituted with other, more heavy, monovalent ions. The last requirement resulted in the necessity to use the electrolyte not in a solid, but in a gel form. Otherwise, due to the low mobility of more heavy ions in solid electrolytes, the required time of the experiment would be very long. Regarding all other aspects, the

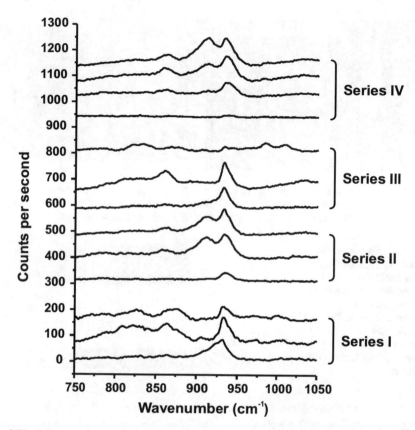

Fig. 2.21 Raman spectra, measured in the polyaniline-polyethylene oxide active heterostructure of the organic memristive device (in the point, near the silver wire electrode) during cyclic application of resisting switching voltages (+0.6 V and −0.2 V) between source-drain electrodes. Series I: −0.2 V was applied; curves from bottom to top: immediately after the voltage application, 15 and 30 minutes after the application; Series II: +0.6 V was applied; curves from bottom to top: after 2, 5, and 30 minutes after the application; Series III: −0.2 V was applied; curves from bottom to top: after 2, 20, and 60 minutes after the application; Series IV: +0.6 V was applied; curves from bottom to top: after 2, 5, 30, and 60 minutes after the application [230]. (Reprinted from Berzina et al. [230], with the permission of AIP Publishing)

used element was absolutely identical to a standard polyaniline-based organic memristive device.

A photograph of the used device and electrical circuit of its connection to the bias and measuring systems are shown in Fig. 2.22 [240].

The measurements were performed in the grazing angle incidence geometry in order to have fluorescence only from the active polyaniline layer, as is shown in Fig. 2.23.

In this case, the working channel, containing 48 molecular layers of polyaniline, was deposited onto a thin film of Kapton (12.5 microns thick). The film of Kapton with deposited polyaniline was placed onto a Teflon sample holder, where two metal

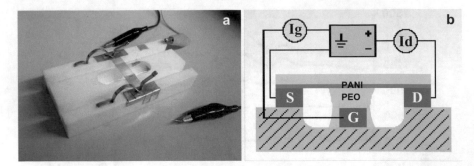

Fig. 2.22 Photograph of the device, used for the X-ray fluorescence experiment (**a**) and the electronic circuit for the switching of the device resistance state (**b**) [240]. (Reprinted with permission from Berzina et al. [240]. Copyright (2009) American Chemical Society)

Fig. 2.23 Scheme of the experiment: grazing angle X-ray excite the fluorescence near the device channel; analyzed fluorescence intensity reveals real-time displacement of ions between PANI channel and electrolyte

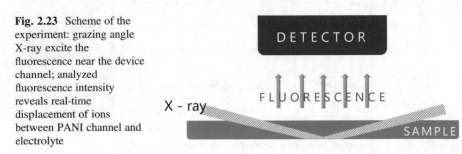

electrodes were placed for making contact with the channel. In the middle part of the sample holder, there was a well, filled with electrolytes, composed of a gel of 0.1 M of RbCl aqueous solution of polyethylene oxide in the concentration of 5 mg/ml. The Kapton was placed over the Teflon sample holder in such a way that the polyaniline channel was in a contact with electrolytes, as is shown in Fig. 2.22. Silver wire was placed at the bottom of the well and was used as a reference electrode. Electrical connections, used for the real-time variations of the device conductivity states, are shown in Fig. 2.22b.

During this experiment, the voltage was applied to the drain electrode, and two current values were measured simultaneously: drain current (total current of the device) and current in the reference electrode circuit (ionic current between polyaniline and electrolyte in the active zone). The measurements were done similarly to the case, described in Sect. 2.3 for the devices on the solid supports and with solid electrolytes. Before the experiments on simultaneous analysis of electrical and X-ray fluorescence characteristics, cyclic voltage-current characteristics of the device, schematically shown in Fig. 2.22, were acquired, and their characteristic features were found to be similar to those obtained for the device on solid supports and with solid electrolyte, indicating the identity of the working principles of these devices.

The excitation of the fluorescence was done using X-ray radiation with an energy of 22.08 keV in the grazing angle geometry. The motion of the monovalent metal

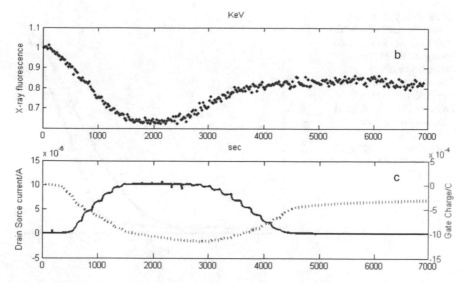

Fig. 2.24 Temporal dependences of the variation of the normalized rubidium fluorescence (upper panel), and drain current and passed ionic charge (lower panel), temporal variation of the voltage, applied to the drain electrode [240]. (Reprinted with permission from Berzina et al. [240]. Copyright (2009) American Chemical Society)

ions (rubidium in this case) was done by registering two characteristic lines (13.39 and 14.96 keV) with a detector, placed at 10 mm upper the active zone of the device. The intensities of the registered lines were normalized to the intensity of elastic scattering for avoiding the effects of the experimental geometry influence. Measurements of the X-ray fluorescence were done simultaneously with the registering of electrical characteristics of these samples.

Figure 2.24 (upper part) reports the temporal dependence of the variation of rubidium fluorescence intensity in the active surface layer. The lower part of Fig. 2.24 shows the temporal variation of the drain current and passed ionic charge (integral of the current in the reference electrode circuit), measured simultaneously with the experiment on the variation of X-ray fluorescence with the same time scale (applied voltage was varied from 0 to +1.2 V, then from +1.2 to −1.2 V, and then back to zero with a step of 0.1 V and time delay of 60 s for each point).

Registered dependence of the charge (integral of the current) variation in the circuit of the reference electrode allows to make an analogy with the ideal memristor of Chua [18]. According to this work, the resistance of the memristor must be a function of the charge that has passed through it. In our case, the situation is slightly different. The conductivity depends not on the total charge, passed through the device (integral of the drain current), but on its ionic component (integral of the current in the reference electrode chain), which is also confirmed by direct measurements of the rubidium ion motion between active zones of polyaniline and electrolyte.

Fig. 2.25 Cyclic voltage-current characteristics of the sample, shown in Fig. 2.22, having the temporal dependences of the fluorescence and electrical parameters, shown in Fig. 2.24 [240]. (Reprinted with permission from Berzina et al. [240]. Copyright (2009) American Chemical Society)

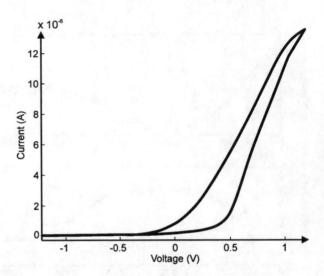

Cyclic voltage-current characteristics of the drain current, corresponding to the results, reported in Fig. 2.24, are shown in Fig. 2.25 (current values for this dependence were taken after 60 s delay after the application of each voltage value).

Initially (low values of the applied voltage), the device is in its low conducting state, corresponding to the high concentration of the rubidium ions in the active zone of the polyaniline channel (upper panel of Fig. 2.24). After a certain value of the applied voltage, polyaniline in the active zone becomes oxidized, which increases significantly its conductivity. This process is accompanied by the repulsion of the rubidium from the polyaniline layer to the electrolyte, which is confirmed by X-ray fluorescence data, reported in Fig. 2.24. The measured value of the rubidium fluorescence intensity reaches its minimum when the device is in its high conducting state. Diminishing the applied voltage value from its maximum to about zero, we can observe a linear decrease of the drain current. Polyaniline in the active zone is in its oxidized conducting state, and rubidium ions do not enter the polyaniline layer. When the applied voltage becomes negative (deviation from the linearity occurs also for small positive values), we see the decrease of the device conductivity, accompanied by the increase of the rubidium fluorescence in the surface layer.

The process of the resistance switching of the device can be described by Formula 2.1:

$$PANI^{+} : Cl^{-} + Rb^{+} + e^{-} \Leftrightarrow PANI + RbCl \qquad (2.1)$$

The transfer of polyaniline in the active zone from its conducting to the insulating state requires the attachment of the electron to the polymeric chain, which is accompanied by the rubidium entrance into the layer, for electrostatic compensation of the charge, induced during doping of the polymer (in the simplest case – Cl^{-}). The

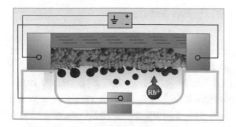 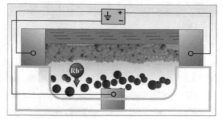

Fig. 2.26 Scheme of processes, occurring in organic memristive devices when negative (left part) or positive (right part) voltage is applied to the drain electrode [240]. (Reprinted with permission from Berzina et al. [240]. Copyright (2009) American Chemical Society)

process is a reversible one, and polyaniline in the active zone (and the entire device) can be transferred back from its insulating state to a conducting one.

Thus, Chua's equations for the memristor can be rewritten for the case of polyaniline memristive device in the following way (Formula 2.2):

$$V = R(w)i_{tot}$$
$$\frac{dw}{dt} = i_{ion}$$

(2.2)

The only difference of the dependences, expressed by Formula 2.2, from those for the ideal memristor, is the fact that the resistance is a function not of the total charge, passed through the device, but only of its ionic component.

The mechanism, schematically illustrating the resisting switching process of organic memristive devices, is shown in Fig. 2.26.

Summarizing, the simultaneous measurements of electrical and X-ray fluorescence intensity characteristics of the device allowed to determine a mechanism responsible for the resistance switching. It has been shown that the mechanism implies the motion of metal ions between active zones of a polyaniline layer and an electrolyte. This work [240] was the first one that has directly demonstrated that the resistance state of memristive devices is a function of one component of the charge that has passed through it.

These results, together with the results of spectroscopic studies, presented in the previous section, allow to conclude that similar mechanisms are responsible for the resistance switching also in devices, using solid electrolytes. It is interesting to note that characteristic time values were similar for devices, using electrolytes in solid and gel states. Probably, it is connected to the fact that a significant difference in the ionic radii of lithium and rubidium compensates for the difference of mobility of these ions in different media.

2.5   Electrical Characteristics in a Pulse Mode

All the results, reported up to this section, were obtained in DC mode. The choice of this mode seems to be absolutely adequate for studying the device that would be used in analog adaptive systems, capable of learning. However, in order to build systems, mimicking processes in living beings, it is necessary to work in a pulse mode, as the signals in nervous systems are propagated in form of spikes [241], which is determined by materials and processes in animals and humans. Therefore, this section will be dedicated to the study of the device characteristics, when input signals are in a form of pulses [242]. The discussion of the realization of adaptive systems, based on these devices, in a pulse mode, will be presented in successive sections.

Devices used for the investigation in pulse mode were identical to those discussed in the previous section for the DC mode.

Pulses of rectangular shape were used for this study, as is shown in Fig. 2.27.

It is very important to have a baseline of the input voltage pulses different from zero, as is shown in Fig. 2.27. As was described in previous sections, the working principle of the device is based on redox reactions. Therefore, the baseline level must have such a value that no redox reactions can occur at this potential level, and the conductivity state must not be changed when no pulse is applied.

Values of the oxidizing and reduction potentials for bulk polyaniline are about +0.3 V and +0.1 V, respectively. However, in devices, the resistance switching

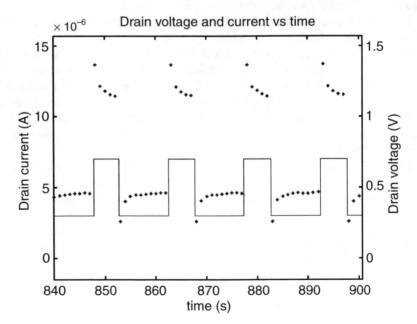

Fig. 2.27 The shape of the input voltage pulses and characteristic current outputs [242]. (Reprinted from Smerieri et al. [242], with the permission of AIP Publishing)

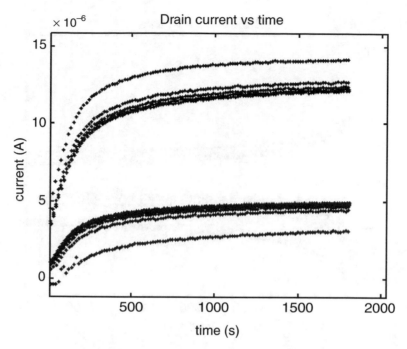

Fig. 2.28 The variation of the drain current of organic memristive devices as a result of the application of rectangular pulses of the voltage. The amplitude of pulses was +0.4 V with respect to the baseline level of +0.3 V. Pulse duration was 5 s and interval of two successive pulses was 10 s [242]. (Reprinted from Smerieri et al. [242], with the permission of AIP Publishing)

occurs at different values of the potentials, due to the fact that the applied voltage is distributed along the whole channel length. Thus, the value of the baseline of the voltage when the resistance switching of the device state does not occur is about +0.3 − +0.4 V.

Characteristics obtained when positive voltage pulses were applied to the device are shown in Fig. 2.28.

The characteristics shown in Fig. 2.28 are significantly different with respect to similar characteristics obtained in DC mode. The reason is connected to the fact that here, we must consider also transition processes that resulted from the presence of the capacitor in the system. Therefore, the kinetics of the conductivity switching is due not only to the ionic drift but also to the RC constant of the system.

The characteristic appears to include two groups of several different lines because, as it is clearly shown in Fig. 2.27, the typical response of our device to a voltage variation is a current that decays from an initial value to a lower, more stable one. Such behavior is determined by the RC of the circuit containing also a capacitance due to the presence of the electrolyte layer. Thus, the more meaningful current values are those taken just before the next voltage pulse application. In Fig. 2.28, the values taken at the end of the voltage pulse correspond to the lowest

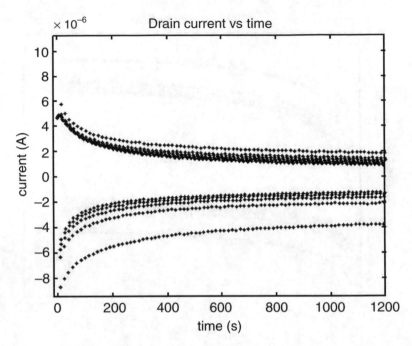

Fig. 2.29 The variation of the drain current of organic memristive devices as a result of the application of rectangular pulses of the voltage. The amplitude of pulses was −0.6 V with respect to the baseline level of +0.3 V. Pulse duration was 5 s and the interval between two successive pulses was 10 s. (Reprinted from Smerieri et al. [242], with the permission of AIP Publishing)

line of the upper group, while those taken just before the pulses correspond to the topmost line of the lower group.

As it is clear from Fig. 2.28, the conductivity of the device was significantly increased after about 30 min (about 120 pulses) both for the applied background voltage level (the value of current was increased from 0.965 μA to 4.89 μA) and for the applied max voltage level (the value of current was increased from 3.04 μA to 12.15 μA). It is to note that the general behavior and characteristic time are very similar to the characteristics observed in DC mode (Fig. 2.10a).

Let us consider now what will happen if pulses of opposite polarity will be applied. As our baseline level is +0.3 V, the pulses of −0.6 V were applied in order to have a negative voltage value of −0.3 V during the pulse application. The resultant characteristics are shown in Fig. 2.29.

As it can be seen from Fig. 2.29, the value of the current, corresponding to the baseline value of the applied voltage, was decreased from 4.93 μA to 0.97 μA, and the value, corresponding to the peak of the voltage, was varied from 4.9 μA to 1.22 μA. As in this case the voltage difference between baseline and peak values is higher with respect to the previous case, the contribution of circuit RC in the transfer kinetics is more pronounced.

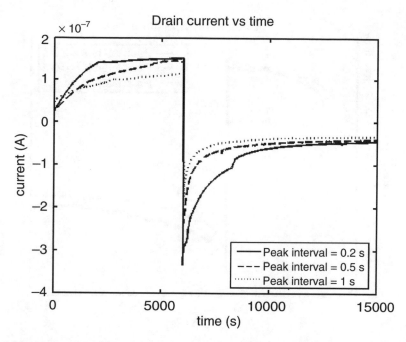

Fig. 2.30 Temporal variations of the total current of the device in a pulse mode. The values were measured at the end of each applied pulse. The time between pulses is equal to the pulse duration. (Reprinted from Smerieri et al. [242], with the permission of AIP Publishing)

It is interesting to note a significant difference in the resistance switching kinetics in DC and pulse modes. In the case of DC, these processes were significantly different: switching from high to low conductivity states was much faster than from low to high ones. Instead, in the pulse mode, time constants are comparable. For example, if we consider a time when the current reaches 90% of its saturation value, we have about 600 s for positive pulses and 800 s for negative ones. This behavior can be explained by the fact that the baseline level in this case is different from zero. Therefore, during the application of the positive pulses, the entire area of the active zone is at the oxidizing potential and transfer simultaneously into the conducting state (in the DC case, simultaneous transition was possible only in the case of the negative applied voltage).

For constructing functional circuits and systems, based on these elements, it is important to study the effects of the pulse duration and time interval between successive pulses on the final characteristics of devices. Results of experiments, carried out to understand these effects, are shown in Figs. 2.30, 2.31, and 2.32.

For all these experiments, a sequence of positive pulses was applied for 10 min, followed by a sequence of negative pulses for 15 min. A DC negative voltage was applied to the devices for 5 min before these experiments for bringing the devices to the initial insulating state.

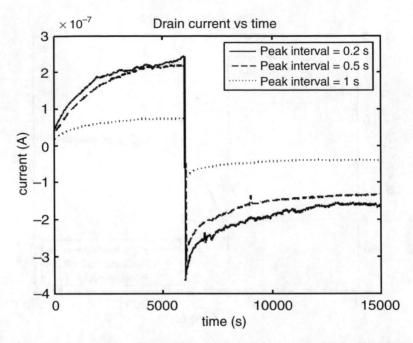

Fig. 2.31 Temporal variations of the total current of the device in a pulse mode. The values were measured at the end of each applied pulse. The time between pulses is triple the pulse duration. (Reprinted from Smerieri et al. [242], with the permission of AIP Publishing)

As it is clear from Fig. 2.32, shorter intervals between pulses result in a more effective variation of the device resistance state.

The results presented in this section demonstrate that the device works effectively also in the pulse mode. As it has been shown, the pulse duration is not a critical parameter, affecting the conductivity variation. The ratio between the time between pulses and the pulse duration seems more important. This conclusion will be later used in a chapter dedicated to the neuromorphic applications of these devices.

2.6   Optimization of Properties and Stability of the Device

Similar to all electronic devices, stability is a very critical parameter also for organic memristive elements. Therefore, a special part of the study was dedicated to the investigation of the device stability and establishing of methods of its improvement. This part will contain three subsections: the stability of the device, optimization of the PANI channel of the device, and role of the electrolyte on the device properties.

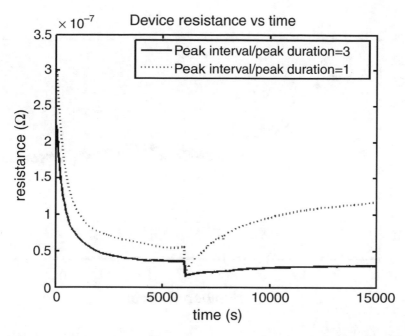

Fig. 2.32 Temporal dependences of the device resistance for two ratios of the time distance between pulses to the pulse duration. For both cases, the pulse duration was 0.5 s. (Reprinted from Smerieri et al. [242], with the permission of AIP Publishing)

2.6.1 Stability of Organic Memristive Device Properties

Before studying the stability of the device properties, the stability of the conductivity of the PANI layers used for the formation of the device channel was studied. The realized layers of used commercial PANI had a conductivity of about 30 S/cm, which is lower than the best-published values [243, 244]. However, these values allowed to register reliable cyclic voltage-current characteristics with a low signal-to-noise ratio.

It was found that the stability of PANI layers was rather high: the conductivity was reduced by less than 1% during the application of 1 V for 48 hours [245].

All components of the current (electronic and ionic) were considered during the study of the device characteristics. However, the most important one for the design of the circuits and networks is the total current passing through the device.

Before studying the "stability," it is necessary to define this parameter because the characteristic property of the device is the capability to change its electrical properties according to the actual combination of input stimuli, their duration and frequency of repetition, and past involvement of the devices into the signal transmission.

For the characterization of the stability of device properties, it seems useful to consider the maximum current value, which can be reached during the application of

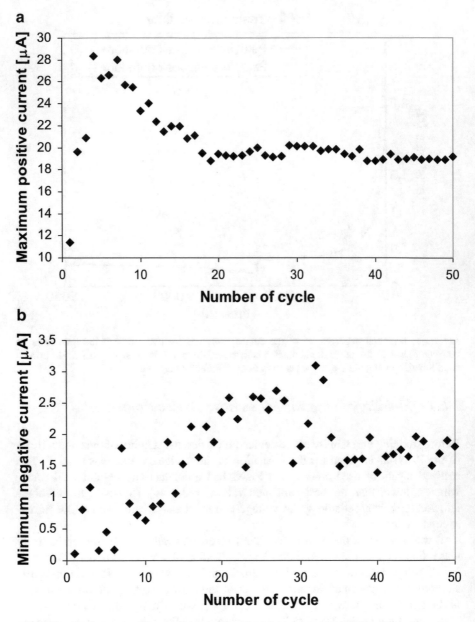

Fig. 2.33 Dependences of the maximal positive (**a**) and minimal negative (**b**) values of current on the number of cycles of voltage-current characteristics. (Reprinted from Erokhin et al. [245], with permission from Elsevier)

positive voltages; the minimum current value, reached during the application of negative voltage; and the ratio of conductivities in ON and OFF states.

Dependences of the first two mentioned characteristics on the number of the cycle of voltage-current characteristic (up to the 50th cycle) are shown in Fig. 2.33.

The observed increase of the maximal positive value of the current during the first four cycles is connected to the described effect of the improvement of the contact between PANI layers with chromium electrodes [246]. Successive 10 cycles result in an about 20% reduction of the maximal current. After it, the value is stabilized with a small tendency to the reduction.

In the case of negative values of the voltage and current, the dependence shows the increase of the conductivity for the first four cycles that is also connected to the improvement of the contact between PANI and electrodes [246]. After it, we can see a rather random scattering of the current values. This fact indicates that the aging of the resistance properties is more pronounced for the conductive state of PANI.

For practical reasons, it seems useful to estimate whether the reported stability is enough for the training of networks, composed of such elements. For the network composed of eight elements (will be discussed later in the section dedicated to the network training), it was necessary to apply 5 min action for the conductivity suppression and about 20 min for its reinforcement [247].

The duration of each cycle, shown in Fig. 2.33, was 80 min. Therefore, the shown 50 cycles correspond to 4000 minutes [248]. We can expect that the network composed of such elements will allow at least 800 training acts. In reality, this value can significantly increase taking into account that the ON/OFF ratio of 10 will be already enough for the training of the system.

The results presented in this section have demonstrated that the stability of properties of single devices is maintained within at least 4000 minutes. Very likely, the factor responsible for the degradation of electrical properties is connected to the dedoping of PANI layers due to the small size of the doping agent (HCl).

The next section will be dedicated to the search of methods for the minimization of this effect.

2.6.2 Optimization of the Device Architecture

Realization of the neuromorphic network with brain-like information processing principles requires a system with a huge number of synapse-like elements working for a rather long time, necessary for effective learning [248] (for example, the human brain contains about 10^{15} synapses). Therefore, individual network elements (memristive devices) must correspond to several essential requirements.

First, the ON/OFF conductivity ratio of the device must be as high as possible. For the devices described in previous sections, this ratio was about 100.

Second, the absolute value of the device conductivity, when PANI of the active zone is in the oxidized state, must also be as high as possible. In fact, if signal pathways of such networks will contain thousands of elements, the high conductivity of each of them will guarantee a high signal-to-noise ratio of the whole system.

Finally, the stability of each element is also a very important parameter. Similarly to the previous section, we assume here that stability means the capacity of the

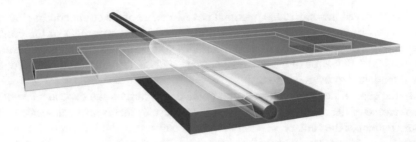

Fig. 2.34 Scheme of the organic memristive device with a "sandwich" structure. (Reprinted from Berzina et al. [249], with the permission of AIP Publishing)

device to change its electrical properties in a reproducible way from one cycle to the other.

In this section, we will consider the effect of the composition and doping of the PANI channel on the properties of the memristive device with a special emphasis on the three characteristics listed above.

Considering dedoping as a main source of instability, two strategies, going in parallel, were carried out in this section of the work for preventing this negative effect [249].

First, the architecture of the device was modified for preventing contact of the active zone with the environmental atmosphere (mainly, for minimizing the effect of the humidity). Scheme of the modified organic memristive device is shown in Fig. 2.34.

The assembling procedure of the device, shown in Fig. 2.34, was done in the following way. Conducting PANI channel was formed on the supports with two metal electrodes similarly to devices, described in previous sections. A groove with a depth, corresponding to the silver wire diameter, was formed in the other solid support (glass). Silver wire was placed into this groove and covered with electrolyte, based on polyethylene oxide with lithium salt. After the electrolyte drying, both supports were mechanically contacted as is shown in Fig. 2.34.

The second strategy was based on the use of polyaniline, doped with heavy agents, inserted into the material during the synthesis stage. For this reason, dodecyl benzene sulfonic acid (DBSA) was used.

The synthesis process is described in [250, 251]. Briefly, in a two-necked flask equipped with a mechanical stirrer and a dropping funnel, 14.40 g of DBSA, dissolved in 800 ml of water, and 4 g of aniline were gently added. The mixture was stirred for 3 hours at room temperature until the formation of the typical milk-like solution of the anilinium salt and then the solution was cooled till 0 °C by dipping the flask into a water-ice bath. Successively, five drops of a saturated solution of cobalt sulfate were added to the anilinium salt solution, and then 10 g of ammonium persulfate, dissolved in 35 ml of water, was slowly dropped to the solution. After 5 hours, the stirring was stopped, and the solution containing a blue-dark precipitate was stored for a night at room temperature. The precipitate was collected by adding 1 l of methanol to the suspension and by filtering it with a

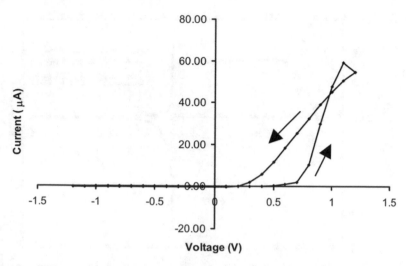

Fig. 2.35 Cyclic voltage-current characteristic of organic memristive devices with "sandwich" architecture and with a channel of polyaniline, containing DBSA molecules, incorporated at the synthesis stage. Arrows indicate the direction of the applied voltage variation. (Reprinted from Berzina et al. [249], with the permission of AIP Publishing)

Buchner filter. Finally, the solid powder was repeatedly washed with methanol and water until neutrality. The PANI was then dried for several days under a vacuum pump.

Cyclic voltage-current characteristics of the device with a "sandwich" architecture and with a channel, realized from polyaniline, prepared by the method, described above, are shown in Fig. 2.35.

The characteristics shown in Fig. 2.35 demonstrate both the hysteresis and rectification behavior. The values of the potentials corresponding to the switching from the insulating form to the conducting one and vice versa are similar to those for devices where the active channel was fabricated from polyaniline with its successive doping with hydrochloric acid.

The analysis of the characteristics allowed to conclude that the performed modification results in the meeting of two requirements listed above. In fact, the device has very high conductivity (60 µA at 1 V with respect to 1 µA at 1 V in the case of devices, described in previous sections). Even more important, the ON/OFF ratio in this case was found to be 2000 that is more than one order of magnitude better with respect to devices made by traditional methods.

Kinetics of the conductivity variation is also very important for the realization of adaptive networks. Temporal dependences of ionic and electronic currents in the device, when positive and negative potentials are applied, are shown in Fig. 2.36.

If we compare these results with those presented in the previous sections, we can claim that characteristic times of the resistance switching in the case of negative applied voltage are the same, while for positive bias voltages, the process becomes significantly faster. This behavior can be connected to the increase of the ON/OFF

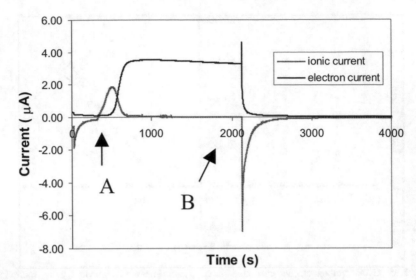

Fig. 2.36 Temporal dependences of the variation of ionic end electronic currents of the device, shown in Fig. 2.34, when positive (+0.6 V, point A) and negative (−0.2 V, point B) potentials are applied. (Reprinted from Berzina et al. [249], with the permission of AIP Publishing)

ratio. In fact, according to the qualitative model of the process presented in Sect. 2.3, the slow process of the device transfer into a conducting state was attributed to the continuous shift of oxidized PANI areas in the active channel in the direction from source to drain. In the case of increased ON/OFF ratio, which we have for the device, shown in Fig. 2.34, these transitions must occur significantly faster. Anticipating the discussion of the models, describing quantitatively the device, I want to mention that these findings are in good agreement with the prediction of time constants for devices with different ON/OFF ratios.

The dependence of the device characteristics on the number of the cycle of voltage-current characteristics is shown in Fig. 2.37. In this case, a standard variation of the applied voltage was used: starting point – 0 V, increasing of the applied voltage with a step of 0.1 V till +1.2 V with a delay of 60 seconds between the voltage application and readout of the current value, decreasing of the applied voltage with a step of 0.1 V till −1.2 V with a delay of 60 seconds between the voltage application and readout of the current value, and increasing of the applied voltage with a step of 0.1 V till 0 V with a delay of 60 seconds between the voltage application and readout of the current value.

Despite the fact that during initial cycles we can see some degradation of conducting properties, after several cycles, the device properties are stable, and all three characteristics, underlined at the beginning of the section, correspond to the criterium, allowing the realization of adaptive analog networks. To illustrate the statement, let us consider cyclic voltage-current characteristics corresponding to the 9th, 10th, and 11th cycles, measured on this device. These characteristics are shown in Fig. 2.38.

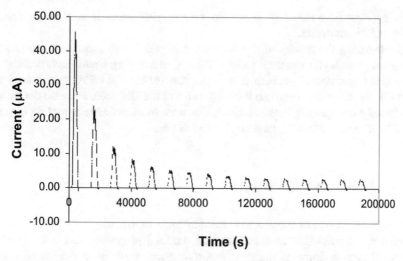

Fig. 2.37 Temporal variation of the total current of the device on the number of the cycle during voltage-current characteristics acquisition. (Reprinted from Berzina et al. [249], with the permission of AIP Publishing)

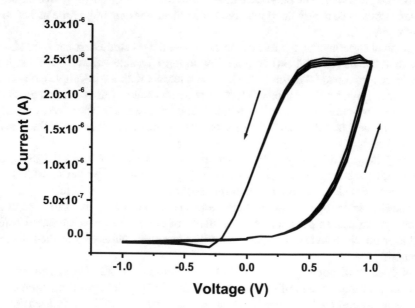

Fig. 2.38 Cyclic voltage-current characteristics of the optimized organic memristive device, corresponding to the 9th, 10th, and 11th cycles. (Reprinted from Berzina et al. [249], with the permission of AIP Publishing)

As it is clear from Fig. 2.38, the device characteristics remain stable after a certain number of initial cycles.

Summarizing the results of this section, it is possible to claim that the use of a strong acid with high molecular weight (DBSA) as a doping agent, incorporated into PANI at the synthesis stage, results in the increase of the ON/OFF ratio of the device conductivity, increase of the conductivity value in the ON state, and also decrease of significantly dedoping effects, which can be even more pronounced using a special "sandwich" architecture of the device construction.

2.6.3 Role of the Electrolyte

The previous section of the chapter was dedicated to the role of the active channel material and architecture of the system on the final properties and stability of the device. However, there is another important component of the element – solid electrolyte – that can also affect the final properties and stability [252].

This section will be dedicated to the role of the composition of the electrolyte on the device properties. The polymeric matrix will be the same, polyethylene oxide, as it was shown to be a very good material for applications demanding the use of solid electrolytes.

In initial experiments, the role of the chain length was studied. It has been shown that short chains (12–35 kDa) do not allow to form a stable gel even at rather high concentrations (up to 60 mg/ml) [252]. For this reason, with these samples, it was not possible to form a stable stripe of electrolyte in the center of the active channel: the deposited drop tended to cover the whole PANI layer surface. Therefore, only high molecular weight fractions of polyethylene oxide were used for the device fabrication.

The following lithium salts were used as doping agents: $LiClO_4$, $LiCF_3SO_3$, and $LiBF_4$. LiCl was also checked, but due to the high hygroscopic properties of this salt, it was not possible to form a stripe of the electrolyte in the solid form.

The electrical properties of elements were analyzed just after the formation of the electrolyte stripes. In particular, the resistance of the devices was measured before and after the electrolyte deposition, as well as after the connection of the reference electrode.

In the case of doping with $LiCF_3SO_3$, the resistance of the PANI channel was increased for three orders of magnitude after the electrolyte deposition, similarly to the situation observed previously for all devices, where $LiClO_4$ was used for the solid electrolyte formation. The situation was absolutely different in the case of using $LiBF_4$ – the resistance of the channel remained practically the same after the electrolyte deposition. In all experiments, when the electrolyte contained this compound, the conductivity reduction was less than 1 percent from its initial value. This behavior is due to the slightly acid nature of $LiBF_4$. Therefore, the deprotonation process is less probable, and the additional doping of the active channel is not required in this case. The last feature seems very important for the applications

Fig. 2.39 SEM image of
48 layers of polyaniline.
(Reprinted from Berzina
et al. [252], Copyright
(2010), with permission
from Elsevier)

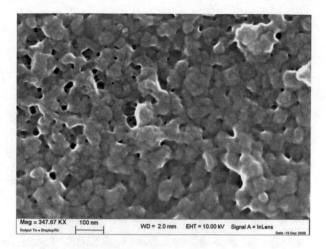

connected to the formation of stochastic self-assembled networks (secondary doping
can vary significantly the formed structure) and to the interaction with living beings
and cells (secondary doping with acid can kill cells and living beings).

The morphology of the polyaniline channel in a contact with solid electrolyte is
shown in Fig. 2.39.

As it is clear from Fig. 2.39, the layer morphology is not a continuous one. It
seems that such structure is formed already at the air-water interface, because
polyaniline is not an amphiphilic molecule, as it was confirmed also by direct
experiments, using X-ray reflectometry at the water surface [240]. It is worth
mentioning that such developed morphology is rather an advantage than a drawback.
In fact, the developed surfaces guarantee a larger area of polyaniline-polyethylene
oxide contact where all processes responsible for the resistance switching occur.

Corroborative evidence on the different effects of the ionic dopants on the
electrical behavior of the device has been obtained by FTIR spectroscopy, by
investigating the interactions between the active PANI layer and PEO doped with
different lithium salts. In particular, we have obtained the spectra of pure and doped
PEO as well as conducting and nonconducting PANI separately and in the
PEO-PANI heterojunctions which characterize our memristor. This allowed us to
identify and separate the spectral contributions of the different substances. We found
that most of the spectral features in the heterojunctions were stipulated by the
changes in PANI spectra, with the exception of a strong peak at 1100 cm^{-1}, which
is the strongest peak in the pure PEO sample. Thus, we could identify the effects of
the different PEO dopants on the electronic states of PANI simply by looking at the
FTIR spectra of PANI directly treated with $LiClO_4$ and $LiBF_4$, respectively. Some of
the relevant data are reported in Fig. 2.40. Measured spectra were divided into three
groups and are presented as separate figures in order to avoid confusion. In
Fig. 2.40a, we show the attenuated total reflection (ATR) FTIR spectra of PANI
samples in the conducting (solid line) and nonconducting (dashed line) states.
Figure 2.40b reports ATR FTIR spectra of conducting PANI before (dashed line)

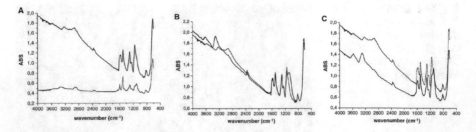

Fig. 2.40 ATR FTIR spectra of PANI films: (**a**) PANI film in insulating (dashed line) and conducting (solid line) states; (**b**) conducting PANI film before (dashed line) and after (solid line) treatment with LiBF$_4$; (**c**) conducting PANI film before (dashed line) and after (solid line) treatment with LiClO4. (Reprinted from Berzina et al. [252], Copyright (2010), with permission from Elsevier)

and after (solid line) its treatment with LiBF$_4$. Figure 2.40c, instead, reports ATR FTIR spectra of conducting PANI before (dashed line) and after (solid line) its treatment with LiClO$_4$. Note the coincidence of the spectra of conducting pure PANI with LiBF$_4$-doped PANI in the critical range 1500–1600 cm^{-1}, due to aromatic ring models and connected to the quinoid-benzenoid ratio in the PANI chain [216]. We also note the similar strength of the peak at 1150 cm^{-1}, which has been associated with high electrical conductivity and a high degree of electron delocalization [216] and hence also correlated to the PANI oxidized state. Although the corresponding spectra for the case of LiClO$_4$ are similar, we note the diminution of oscillator strength of the peak at 1150 cm^{-1}, accompanying the overall decrease of intensity of the Drude-like continuum, which is a signature of the sample metal-like conductivity. The data corresponds well to those obtained with micro-Raman characterization, where the transformation of PANI into the insulating state was observed immediately after the formation of the PEO layer, doped with LiClO$_4$ [230]. Similar but less clear results were obtained in the PANI-PEO heterojunctions, demonstrating directly that the general effect of the ionic dopants on the device behavior is due to specific changes induced by them in the electronic states of PANI connected to the benzenoid and quinoid ring vibrations, i.e., the oxidation and charge transfer states.

For the network realization, it is more important the comparison of voltage-current characteristics of devices, doped with the use of different salts for the electrolyte realization. The characteristics for all three used salts are shown in Fig. 2.41.

All characteristics shown in Fig. 2.41 demonstrate characteristic features of organic memristive devices: rectifying behavior and the presence of the hysteresis. However, the comparison of these characteristics allows to select one of them, namely, the case of electrolyte, containing LiBF$_4$, when a significant shift of the value of the reduction potential was observed, in the case of the device switching to the insulating state. The voltage value was shifted to the negative branch.

This result has particular importance for the design of circuits and networks that must work in the pulse mode and/or mimic the function of living beings' nervous systems. As was shown in the section, dedicated to the working of the organic

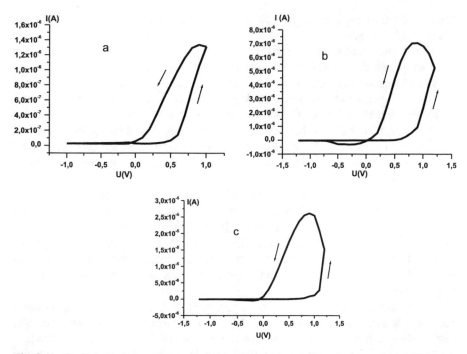

Fig. 2.41 Cyclic voltage-current characteristics of organic memristive devices, where LiClO$_4$ (**a**), LiCF$_3$SO$_3$ (**b**), and LiBF$_4$ (**c**) were used for the electrolyte formation. (Reprinted from Berzina et al. [252], Copyright (2010), with permission from Elsevier)

memristor devices in a pulse mode, an application of a constant offset voltage was required, preventing the device from switching to the OFF state. The characteristics shown in Fig. 2.41c allow to avoid the application of the mentioned offset, as zero potential in this case does not change the conductivity state of the device.

Summarizing, the results presented in these sections give ideas of how the materials of the electrolyte layer influence the properties and stability of the devices.

2.7 Organic Memristive Devices with Channels, Formed by Layer-by-Layer Technique

Most of the results presented in this book were obtained using organic memristive devices, where the channel was formed by the Langmuir-Schaefer technique. The choice of this method was determined by two requirements: on the one hand, the channel must guarantee a high device conductance; on the other hand, it must be thin enough, as the effect of the resistance switching is based on diffusion processes (thick layers will decrease significantly the speed of the resistance switching). It has been established experimentally that the optimal channel thickness is within the range of 25–100 nm.

During this deposition, a solution of polyaniline is spread at the air-water interface. After the spreading, the monolayer is compressed with a barrier till the target surface pressure (10 mN/m). After the compression, the monolayer is split by a special grid with windows, corresponding to the sample sizes. The layer is successively transferred to the sample by touching it horizontally in each window.

However, this technique has several disadvantages, such as the slow deposition process, the necessity of the special equipment, and, probably the most important, the possibility of the channel formation only on flat supports with specially defined physicochemical properties.

There is another method called polyelectrolyte self-assembling or "layer-by-layer" (LbL) deposition that allows also film formation with nanometer resolution [253]. The method is based on the successive deposition of layers of polyanion and polycation molecules from solutions. This technique is very simple and does not require the use of special equipment. Its advantage is that it allows the formation of layers with molecular thickness on any kind of supports disregarding their shape and physicochemical properties.

The application of this technique for the formation of the conducting channel required the use of water solutions of polyaniline that is normally not water-soluble. For making such solution, the technique reported by M. Rubner was used [254].

As the first layer, polyethylene imine was deposited as this compound provides very high adhesion to practically all surfaces. After it, alternative layers were deposited from the solution of polyaniline, prepared according to [254], and polystyrene sulfonate (PSS). The procedure was repeated till the desired thickness of the channel was reached.

The scheme of the device and its connection to the measuring circuit were similar to the above-described cases.

Typical sizes of the fabricated devices were: channel length 7 mm and channel width 3 mm; thickness of the channel varied in the range 2–30 nm.

Solid electrolyte layers and the connection of the reference electrode were done similarly to the previously described cases.

Before studying the device characteristics, the conductance of the PANI-PSS layers was investigated [255]. Voltage-current characteristics of the film, containing 14 PANI-PSS bilayers, before and after doping with hydrochloric acid, are shown in Fig. 2.42.

Both characteristics shown in Fig. 2.42 have a linear character. In contrary to the layers, formed by the Langmuir-Schaeffer technique, we can see a high conductance even without doping. It can be explained by the fact that PANI molecules are in a contact with PSS molecules, acting as doping agents due to their acidic nature. Therefore, the conducting layer is formed already at the film formation phase. However, doping results in the increase of the conductivity that, related to the PANI layers only, is 0.36 S/cm for the undoped samples and 0.83 S/cm for doped samples.

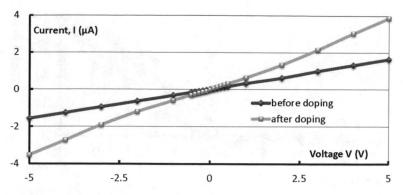

Fig. 2.42 Voltage-current characteristics of the LbL film, containing 14 PANI-PSS bilayers, before and after doping with hydrochloric acid. (Reprinted by permission from Springer Nature, Erokhina et al. [255], Copyright (2015))

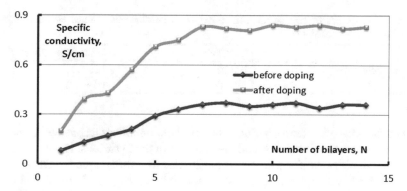

Fig. 2.43 Dependence of the conductivity of the film on the number of PANI-PSS bilayers before and after the doping. (Reprinted by permission from Springer Nature, Erokhina et al. [255], Copyright (2015))

The dependences of the conductivity of these layers on the number of deposited PANI-PSS bilayers are shown in Fig. 2.43.

Similarly to the case of using the Langmuir-Schaeffer deposition, the conductivity value reaches a saturation only when some critical thickness of the layer was formed. The increase of the conductivity on the number of layers before reaching this critical thickness is connected to the inhomogeneous nature of the formed layers due to the interactions with the support material. When the critical thickness was achieved, the support surface becomes to be homogeneous regarding its physico-chemical properties, which results in the reproducible deposition of successive layers with the same electrical properties.

After studying the properties of PANI-PSS layers, characteristics of the memristive devices, based on them, were studied. Cyclic voltage-current characteristics of such device are shown in Fig. 2.44.

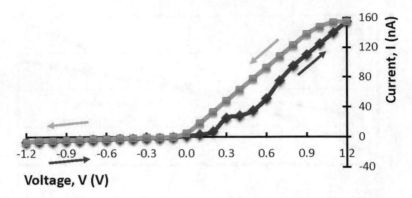

Fig. 2.44 Cyclic voltage-current characteristics of the organic memristive device with the channel, fabricated with LbL technique. Arrows indicate the direction of the variation of the applied voltage. (Reprinted by permission from Springer Nature, Erokhina et al. [255], Copyright (2015))

Table 2.1 Properties of organic memristive devices with channels, formed by the Langmuir-Schaeffer and LbL techniques

	ON/OFF ratio	Max conductivity (S/cm)
Langmuir-Schaeffer	100,000	10–30
LbL	100	0.9–1.2

The characteristics shown in Fig. 2.44 have both important features of the organic memristive device: rectification and the presence of the hysteresis. However, we can see also a significant difference: the voltage value for switching to the ON state was found to be lower and corresponded to approximately +0.4 V. Very likely, it is connected to the presence of the PSS layer that, as it was mentioned above, acts as a doping agent. ON/OFF ratio in this case was found to be about 20.

Comparison of electrical properties of organic memristive devices with channels, formed by the Langmuir-Schaeffer and LbL techniques, is presented in Table 2.1.

As it is clear from the data presented in Table 2.1, characteristics of memristive devices with a channel formed by the LbL technique are worth than those for the Langmuir-Schaeffer method. Nevertheless, even such characteristics allow to form working devices. Moreover, there are additional parameters allowing to consider this method as a very perspective one for the practical realization of organic memristive devices. Table 2.2 reports a comparison of some of these parameters.

Data reported in Table 2.2 show that LbL is a preferential technique regarding all parameters connected to the technological processes.

It is to underline one other important advantage of the method. Coupling of live cells with electronic devices is a very important task, as it allows to study processes occurring in cells and to perform targeted application of external stimuli on desirable areas of grown cell network. The growth of nerve cells on functional electronic circuits is an important task. In this respect, very important approaches were performed by the group of P. Fromherz, where nerve cells were grown on arrays

Table 2.2 Technological parameters of the Langmuir-Schaeffer and LbL techniques

	Restrictions on the support physicochemical and morphological properties	Necessity of doping	Time consumption	Special equipment	Possibility of parallel deposition	Technological compatibility
LS	YES	YES	High	YES	NO	Low
LbL	NO	NO	Low	NO	YES	High

Table 2.3 Conductivity of LbL films with different polymers

Polymer	Conductivity (deposition from water solution) (S/cm)		Conductivity (deposition from NaCl solution) (S/cm)	
	Before doping	*After doping*	*Before doping*	*After doping*
PSS	0.32	1.58	0.38	1.41
P	0.08	1.88	1.32	1.58
SA	0.15	2.43	0.18	1.01

of field-effect transistors [256–260]. Growing activity in the field of memristive devices resulted in the discussion of the possibility of cells growing also on memristive arrays [261]. However, successful utilization of memristive devices (in particular, organic memristive devices) for these applications require two fundamental aspects: biocompatibility of surfaces that will be in a contact with cells; devices must work in liquid media at pH values compatible with physiological ones. It is obvious that channels formed by the Langmuir-Schaeffer technique do not correspond to the mentioned requirements: polyaniline is not biocompatible material, and its immersion into solutions with neutral pH results in the dedoping of the material. In the case of channels formed by the LbL technique, the situation is different. The film in the active channel is an alternation of layers of two different polymers, one of which can be a biocompatible one. The particular deposition of the channel will provide a situation when this end layer will be in a contact with cells. If this polymer will have acid groups, it will provide also additional doping, which will result in the difficulty of the dedoping process.

A study was performed where LbL layers of polyaniline were alternated with PSS, pectin (P), and sodium alginate (SA). Layers of these polymers were deposited from water solutions and from solutions containing 0.5 M of sodium chloride. Doping was done in the hydrochloric acid solution. The comparison of the electrical properties of these layers is presented in Table 2.3.

As it is clear from Table 2.3, the values of the conductivity for films containing PSS and SA are practically the same, whether the layers were deposited from water solution or from sodium chloride solution. Doping of these films results in the increase of their conductivity. The situation is different for films with P molecules. When the deposition was done from pure water solution, doping results in a 20-fold increase of the conductivity, while in the case of deposition from sodium chloride solution, the values of the conductivity are practically the same before and after doping. Significant increase in the conductivity after doping is due to the fact that P has less acidic groups comparing with PSS and SA. In the case of the deposition from sodium chloride solution, Cl^- ions can provide doping by themselves.

As was mentioned above, for contacting with live cells, active channels must maintain their conductivity in physiological solutions. The dependence of the resistance of LbL film containing PANI-PSS, on the pH of the solution, where the measurements were done, is shown in Fig. 2.45.

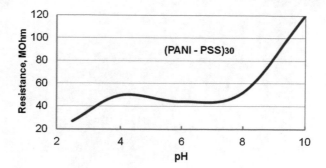

Fig. 2.45 Dependence of the resistance of the LbL film of PANI-PSS on the pH of the solution

As is shown in Fig. 2.45, there is no significant variation of the conductivity in the pH range 4.0–8.0 that corresponds to the physiological solutions, where cell growth occurs.

Summarizing, this section of the book has demonstrated the applicability of the LbL method for the formation of conducting channels of organic memristive devices with properties required for the fabrication of adaptive networks. In addition, the conductivity of such films does not depend on the pH of solutions where the measurements are carried out, which allows the growth of cells on them.

Chapter 3
Oscillators Based on Organic Memristive Devices

In this short chapter, we will discuss the possibility of using organic memristive devices for the realization of systems, capable to generate auto-oscillations.

An important feature of all living beings is the presence of rhythmic periodic processes even in a fixed environment. According to Schrödinger, "life avoids equilibrium" [14].

For the realization of information processing systems, the presence of rhythmic processes is also a necessary requirement. It is to note that this requirement is valid both for traditional computational systems (clock generator) and in the nervous system of living beings (central pattern generator—a system responsible for the spike production). To illustrate the last statement, we can consider a pond snail nervous system: one part of this system is responsible for the generation of rather long sequences of spikes as a result of a single excitation.

Therefore, if we want to use effectively organic memristive devices in neuromorphic networks, we must find a way of the realization of systems, working in an auto-oscillation mode, with minimal modification of the architecture of memristive devices.

Of course, generators of auto-oscillations can be easily fabricated with three-terminals transistor-like elements. However, in order to have a possibility of mimicking synapse properties, organic memristive devices must be used as two-terminal elements with respect to the external circuits. Therefore, the only possibility of the realization of systems working in the auto-oscillation mode is the modification of the organic memristive device structure in such a way that the value of the potential at the reference electrode will be not fixed anymore but will vary during the device function.

The introduction of the element, capable to accumulate charge, into the reference electrode chain seems to be the easiest method for reaching this requirement. The obvious way to do it is to connect an external capacitor into the chain of the reference electrode [262], as is shown in Fig. 3.1.

In the configuration, shown in Fig. 3.1, the presence of the ionic current in an electrolyte of the device will result in charging/discharging of the connected

© Springer Nature Switzerland AG 2022
V. Erokhin, *Fundamentals of Organic Neuromorphic Systems*,
https://doi.org/10.1007/978-3-030-79492-7_3

Fig. 3.1 Schematic representation and connectivity to the external circuit of the organic memristive device for working in auto-oscillation mode [262]. (Republished with permission of IOP Publishing, from Erokhin et al. [262], copyright (2007); permission conveyed through Copyright Clearance Center, Inc.)

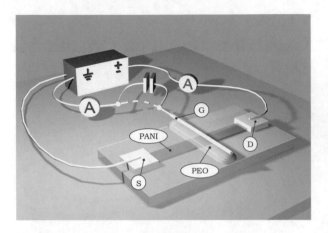

capacitor that will vary also the value of the actual potential at the reference electrode. Therefore, the variation of the PANI channel conductance in the active zone will depend on the actual potential difference of the PANI channel zones with respect to the variable potential of the reference electrode, according to the values of this potential, listed below (Formula 3.1):

$$
\begin{aligned}
&V_< + 0.1 \\
&V + 0.1\ V_< V_< + 0.3\ V \\
&V_> + 0.3\ V
\end{aligned}
\tag{3.1}
$$

It is possible to distinguish three ranges of the potentials of PANI fractions in the active zone with respect to the potential of the reference electrode (Formula 3.1): when the potential is less than +0.1 V, PANI in the active zone is in a condition that allows its switching to the insulating state; when the potential is more than +0.3 V, PANI in the active zone is in a condition allowing its switching to the conducting state; in the range + 0.1 – +0.3 V, the conductivity state of PANI in the active zone is not varied.

Temporal dependences of the experimentally measured values of the current in the drain and reference electrode circuits for the memristive device with attached external capacitor at fixed biased voltage are shown in Fig. 3.2 (a) and (b), respectively.

Figure 3.2 (c) shows combined dependences, shown in (a) and (b) parts.

As it is clear from Fig. 3.2, auto-oscillations can be observed for currents in both circuits, with a phase shift between them. Analysis of the data allows to conclude that the oscillations in the drain circuit are directly connected to the integral of the temporal oscillations in the reference electrode circuit, which is in the agreement with a qualitative explanation of the working mechanism, discussed in Sects. 2.4 and 2.5.

For a qualitative explanation of the oscillation behavior, it is necessary to consider at least two processes, going simultaneously in this system. The first

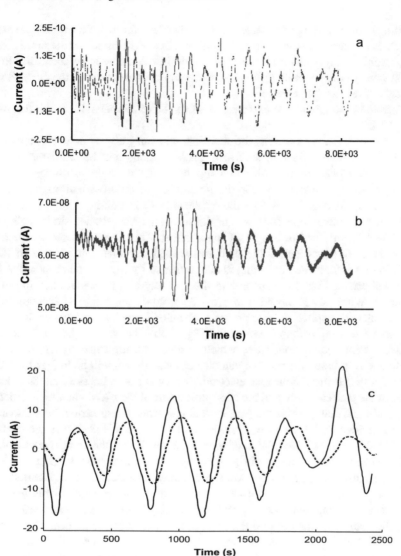

Fig. 3.2 Temporal dependences of the drain (**a**) and reference electrode (**b**) currents for the organic memristive device with the external capacitor of 1.0 μF and the applied external voltage of 1.0 V. Both characteristics are shown in (**c**) for comparison (for better representation, the current values for the reference electrode circuit are multiplied for 100, and the values of current in the drain circuit are shifted for −0.5 nA) [262]. (Republished with permission of IOP Publishing, from Erokhin et al. [262], copyright (2007); permission conveyed through Copyright Clearance Center, Inc.)

process is connected to the motion of ions between the polyaniline channel and solid electrolyte in the active zone, resulted from the actual difference of potentials between the active zone and the reference electrode. The presence of the capacitor in the reference electrode circuit provides a possibility of charge accumulation, resulting in the variation of the actual potential at the reference electrode. The second process is connected to the temporal variation of the distribution of the resistance of the polyaniline layer along the channel length in the active zone according to the actual value of the potential difference of the particular segment of the channel with respect to the reference electrode.

Therefore, the qualitative explanation of the observed behavior is the following one [262].

Let us consider, for example, the situation, when negative voltage was applied to the device in the conducting state (the considerations will be similar if we will consider the application of positive voltage to the device in the insulating state).

The processes, responsible for the generation of auto-oscillations, are shown schematically in Fig. 3.3. Before the application of the drain voltage (Fig. 3.3a), PANI in the active zone is in its insulating form. The applied drain voltage is distributed linearly on the length of the channel. Thus, zones closed to the drain electrode are at the potential, higher than the oxidation one. The process of PANI oxidation starts in these zones that is accompanied by the motion of positive ions (Li + and protons) from the channel to the electrolyte. It results in charging of the capacitor. When zones closed to the drain electrode are oxidized (become more conducting), the applied voltage profile is changed in such a way that central zones also arrive at the oxidation potential (Fig. 3.3b). Thus, they begin to transfer themselves into conducting state, which is again accompanied by the motion of positive ions. It is to note that the velocity of their motion will be higher for zones, just under the position of the gate electrode, which is due to the fact that the distance between the gate electrode and the appropriate part of the PANI channel is orders of magnitude less than that for the peripherical zones and the electric field value is the voltage divided to the distance. Experimental evidence of this fact is presented in Sect. 4.4 of the book. The motion of ions increases further the charge, accumulated on the capacitor and, therefore, the gate voltage. At a certain moment (Fig. 3.3c), the charge, accumulated at the capacitor, significantly increases the gate potential. As a result, some zones of PANI in the active channel are at the reduction potential (actual potential of the zone due to the applied drain voltage minus gate potential). Therefore, they begin to transfer into the insulating state, which is accompanied by the motion of positive ions from the electrolyte to the PANI layer. The capacitor is discharging, and the gate potential is reducing. These processes result in the continuous displacement of the insulating area of the PANI channel in the direction from source to drain (Fig. 3.3d). Finally, practically, the whole channel transfers itself into the insulating state (Fig. 3.3e). The capacitor is discharged, and we have a situation very similar to that, shown in Fig. 3.3a.

Summarizing, the mechanism, responsible for the auto-oscillation generation can be described qualitatively by three simultaneously going processes. First, redistribution of the electrical potential profile occurs on the length of the polyaniline in the active area (the highest component of the potential difference will be concentrated in lower conducting segments). Second, the external capacitor will be charged-discharged according to the actual voltage between the reference electrode and the active area of the polyaniline channel. Third, redistribution of the resistance in the

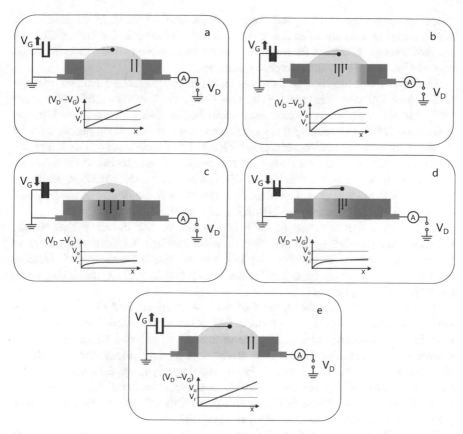

Fig. 3.3 Schematic representation of processes, responsible for the generation of auto-oscillations. The color of PANI in the active zone represents the conductivity state (more green—more conducting). Grey zone represents solid electrolyte. Red level within the capacitor represents the accumulated charge. Red arrow shows whether the gate voltage (resulted from the charging or discharging of the capacitor) increases or decreases. Blue arrows in the PANI active zone indicate the motion of positive ions (Li + and protons). Their length corresponds to the intensity (number and velocity) of this motion. Bottom panels show the distribution of the potential (applied drain voltage minus gate voltage, resulted from charging of the capacitor)

active zone will result in the fact that some segments of the area will be at a potential lower than the reduction one, and the discharge of the capacitor will occur, accompanied by the back flow of positive ions. This ionic flow will vary the resistance profile, which, as a consequence, will vary the potential distribution profile.

As it is clear from Fig. 3.2, oscillations of the current in the circuit of the reference electrode occur around its zero value. Therefore, each moment is characterized by a preferential direction of the ionic flow. Therefore, the increase or decrease of the total device conductivity is connected to the direction of the ion motion in the polyaniline—polyethylene oxide junction. Integral of the current in the reference electrode circuit is zero. The phase shift of the oscillation of the total and ionic currents is connected to the fact that the resistance of the device depends on the transferred ionic charge (temporal integral of the ionic current). In the case of the total current, its average value is not equal to zero, because in this case, we have a directed transfer of carriers from source to drain and not the periodic motion of ions between the active area and reference electrode.

Qualitative consideration of these processes allows to make a parallel with the well-known Belousov-Zhabotinsky reaction [263], the importance of which, especially for the description of processes occurring in living beings, was strongly underlined [264]. For making these cyclic reactions, at least three processes must go simultaneously: oxidation, reduction (at least one of these reactions must be autocatalytic, producing catalyzer during the reaction), and the process of the catalyzer inhibition. In our case, we also have oxidation and reduction reactions, while the variation of the potential at the reference electrode and the profile of the potential distribution along the active area play roles of the catalyzer and inhibitor.

It is to note that the most of published works on the Belousov-Zhabotinsky reaction have reported cyclic variation of color and/or viscoelastic properties of the reaction medium. In our case, instead, we see also the variation of electrical properties, which is very important for the application in devices, mimicking some properties of living beings.

The device, described above, required the use of the external electronic component—capacitor. However, further studies of the organic memristive devices have revealed the possibility of avoiding the utilization of external elements. For this reason, it was necessary to substitute silver wire as a reference electrode with a material that can accumulate ions. Highly oriented pyrolytic graphite was chosen for this reason due to its well-known utilization as an electrode in rechargeable batteries [265]. Charge accumulation is due to the possibility of ion penetration between planes of the crystallographic lattice of the graphite.

The modified device has an architecture very similar to that of the organic memristive device, described in Chap. 2 with the only difference: the reference electrode was fabricated from a narrow strip of the fresh cut of highly oriented pyrolytic graphite. Temporal variations of the total current acquired after the application of different values of constant voltages are shown in Fig. 3.4.

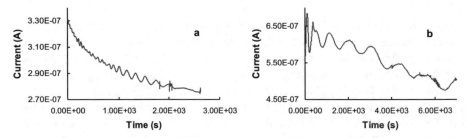

Fig. 3.4 Temporal dependences of the variation of the total current in organic memristive device with the reference electrode, fabricated from highly oriented pyrolytic graphite, acquired during the application of a constant voltage of +5.0 V (**a**) and − 5.0 V (**b**) [262]. (Republished with permission of IOP Publishing, from Erokhin et al. [262], copyright (2007); permission conveyed through Copyright Clearance Center, Inc.)

Summarizing, it was described the architecture and properties of the device capable to generate auto-oscillations of current at fixed applied voltages. This behavior was explained qualitatively by the periodic variation of the polyaniline channel resistance in the active zone, resulted from the continuous variation of the potential profile and cyclic charging/discharging of the capacitor in the reference electrode circuit.

These results establish a basis for the development of models, describing the working principle of the device. These models will be considered in the next chapter.

Chapter 4
Models

The experimental data presented in previous sections required the development of the theoretical model describing all observed phenomena. In particular, it is necessary to explain not only obvious difference in the kinetics of the switching from the conducting to insulating states and vice versa, but also the generation of auto-oscillations.

In this chapter, we will consider three approaches for model construction. The first one considers processes of the variation of the potential distribution profile on the polyaniline channel length in the active zone induced by redox reactions going at different potentials. This was the first developed model describing well most of the results obtained on organic memristive devices. However, due to a large number of parameters, the calculation of processes occurring even in one device is rather time-consuming which makes difficult its utilization for calculating parameters of circuits, containing a significant number of devices. Therefore, the second simplified approach was developed disregarding microscopic processes in different segments of the device active zone but considering only values of the output current, depending on the history of the device functioning. This section was included for a better understanding of the work of networks which will be discussed in successive chapters. Finally, the last developed model is based only on the use of fundamental physical lows without any assumption.

4.1 Phenomenological Model

The model that will be considered in this section is based on several assumptions. First, it was suggested that all processes occurring in the active zone of the contact of polyaniline channel with solid electrolyte, which is responsible for the resistance variation, are due to the redox reactions. Two polyaniline zones, not in direct contact with polyethylene oxide, can be considered as fixed resistors.

© Springer Nature Switzerland AG 2022
V. Erokhin, *Fundamentals of Organic Neuromorphic Systems*,
https://doi.org/10.1007/978-3-030-79492-7_4

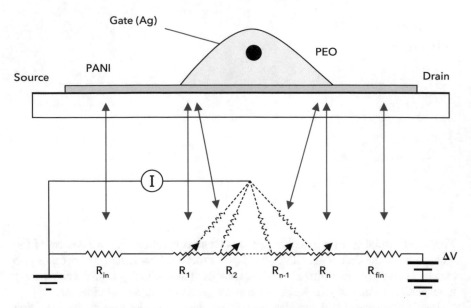

Fig. 4.1 Scheme of the organic memristive device and its equivalent electronic circuit. Variable resistors correspond to areas of polyaniline and polyethylene oxide in the active zone [266]. (Reprinted from Smerieri et al. [266], with the permission of AIP Publishing)

The scheme illustrating the model elements is shown in Fig. 4.1.

The next assumption is that the whole layer of polyaniline in the active zone with the same linear coordinate in the axis from the source to the drain has the same potential. Thus, the only spatial variable is the distance from the source to the segment under the consideration.

In the frame of this model, the whole active zone of the device was split into narrow stripes, supposing that all points of the stripe have the same potential [266]. Thus, we postulate the absence of all potential gradients, as well as simultaneity and uniformity of all processes, occurring inside any stripe. Each stripe has its own resistance. We will call this resistance "drain resistance" to distinguish it from "reference electrode circuit resistance" that will be considered later. The value of this resistance can be changed according to the potential that the corresponding stripe has, with respect to the reference electrode and a time interval when the stripe had a potential higher than the oxidation one, or lower than the reduction one. Temporal dependences of the resistance variation were taken from the fitting of experimental data with exponential functions for processes of the device transfer from the conducting to insulating states because it was supposed that this transformation occurs simultaneously of the whole length of the active zone. In principle, this suggestion is not correct as it will be shown during the consideration of the third model. However, even this model explains rather well the obtained results.

Each stripe is connected to the reference electrode via a resistor (the contact through the solid electrolyte). These resistors have constant and variable components. The value of the constant component of these resistors depends on the length with respect to the position of reference electrode (its value is minimal in the central part and maximal at the ends of the active zone, which is due to the increased relative thickness of polyethylene oxide for lateral PANI stripes with respect to the central ones). The variable component of the resistance is connected to the variation of the ionic content of the electrolyte, resulted from the redox reactions that can increase or decrease the concentration of mobile ions in the solid electrolyte.

Temporal variation of the total current of the device is connected to redox reactions in each stripe that reached a potential higher than the oxidation one or lower than the reduction one with respect to the reference electrode. As was shown in Chap. 2, these processes are accompanied by the motion of ions between polyaniline and polyethylene oxide layers. As it was stated before, experimental data corresponding to the transition of the device from the conducting to the insulating states were used for the model development, as the process was considered uniform on the whole length of the active area. As one of the used assumptions was the uniformity of the reaction kinetics within each stripe, this kinetics (negative bias, switching from the conducting to insulating state) was used for the modeling of both reduction and oxidation processes.

Let us consider what happens with the resistances of stripes in the active zone during increase or decrease of the applied voltage in the case when the device was initially in conducting state (this suggestion was done for definite; similar results can be obtained for the device in the initial insulating state). When the voltage between the defined stripe and the reference electrode reaches a threshold value (+0.1 V), corresponding to the reduction potential, and the voltage applied to the device continues to decrease, the resistance of the stripe begins to increase. Similarly, if the voltage between the defined stripe and the reference electrode reaches the other threshold level (+0.3 V), corresponding to the oxidation potential, and the applied voltage continues to increase, the resistance of the stripe decreases. Within the model, each stripe was virtually equipped by a timer to follow the variation of the conductivity state of each stripe, starting from the moment when its potential was higher than the oxidation one, or lower than the reduction one. Reset of the timer was done at each moment when one of the threshold values of the potential of the stripe was reached (each time of the cycle "0" corresponds to the reaching of the oxidation (and its further increase) or the reduction potential (and its further decrease)), indicating the beginning of the process of the resistance switching.

During the model development, time constants and resistance values in each stripe in the active zone were considered as independent one from the other.

The resistance in the circuit of the reference electrode (the resistance between the stripe of the polyaniline channel in the active zone and the reference electrode) is connected to the ionic current in the solid electrolyte. Values of the resistors in the reference electrode circuit must be also variable. In fact, redox reactions imply, as it was shown in Chap. 2, ion motion between polyaniline and polyethylene oxide layers in the junction. Therefore, the concentration of ions with significant mobility (lithium ions and protons) in the solid electrolyte and, as a consequence, the conductivity of the electrolyte will be changed when their concentration is increased or decreased. In the frame of this model, it was suggested that the variation of the current value in the reference electrode circuit is proportional to the conductivity of polyaniline in the active zone. This suggestion is based on the experimental results obtained using X-ray fluorescence measurements presented in Chap. 2, allowed to determine the mechanism of the resistance switching, that can be rewritten for the polyaniline memristive device in the following form for the case of solid electrolyte (lithium ions instead of rubidium in the case of gel electrolyte):

$$PANI^+ : Cl^- + Li^+ + e^- \Leftrightarrow PANI + LiCl.$$

The model was developed and applied for the explanation of two experimentally obtain characteristics.

In the first case, the constant voltage was applied to the drain electrode, with or without the capacitor in the reference electrode circuit. Discrete-time intervals, usually one second, were used for the calculation of the device parameters during modeling. The modeling procedure was applied for 5000–40,000 intervals with used time discretization.

When the positive voltage was applied, we supposed that all polyaniline stripes in the active zone are in an insulating state at the initial moment. Consequently, when the negative voltage was applied, we supposed that all stripes in the active zone are in a conducting state at the initial moment.

Two subprograms were used during modeling, called respectively "envelope program" and "active program." These programs were written by Dr. Anteo Smerieri during his Ph.D. work preparation.

The "envelope program" performed readout and memorizing of the resistance values of each stripe, calculation and memorizing of the potential distribution profile on the length of the active zone, as well as calculation and recording of the value of the total current and linear component of the current in the reference electrode circuit. When all these operations were done, the "active program" was started. The "active program" has checked whether the threshold conditions (oxidation or reduction potentials) were reached for some stripes, if yes, it attributes time "0" to the timer and initial resistance values for stripes, passed to new conditions, and made calculations of the nonlinear component of the reference electrode circuit current. After it, the "envelope program" memorized values of the current in the reference electrode circuit. These operations were repeated cyclically till reaching a predetermined number of calculation cycles. After the end, the dependences of the total current and current in the reference electrode circuit on time were plotted. In the case of the structure with the capacitor, the dependence of the capacitor charge on time was also plotted.

Physical values used during the model construction were taken from our experimental data or from literature; values of R_{max} and R_{min} were attributed to having a total resistance of the polyaniline channel in the active zone to be 500 kΩ, when all stripes are in the highly conducting state, and 1000 MΩ, when they are in the highly insulating state. The values of the resistance of the polyaniline channel between the source and active zone and between the active zone and drain were 200 kΩ and 100 kΩ, respectively. The constant component of the resistance in the reference electrode circuit was supposed to be equal to 50 MΩ, and the coefficient of the proportionality for the nonlinear component of the current in the reference electrode circuit was supposed to be equal to -6×10^{-13} A s/Ω. Both these values were chosen based on the analysis of our available experimental results. Reduction and oxidation potentials were supposed to be +0.1 V and +0.3 V, respectively. The time constant corresponding to the exponential variation of the resistance value was obtained by the approximation of experimental results, and its value was found to be 10^4 seconds.

Other parameters used for the model development, such as the number of stripes, values of the applied voltage, and the capacitor capacitance were not directly connected to the physical characteristics of our devices. The variation ranges of the applied voltage was $-2.0 - +20.0$ V and of the capacity, it was 0.1–100 μF. The number of stripes in the active zone was 100 – an optimal compromise value for the good spatial resolution and the acceptable calculation time.

One second temporal step was used for all models, except the case when the capacity value was less than 1.0 μF. In these cases, the temporal step was reduced.

The first application of the model was done for explaining the difference in the resistance switching kinetics during the application of positive and negative voltage values (experimental results were considered in Chap. 2). Modeling was done for fixed values of the applied voltage.

Experimental and theoretical dependences of the total current variation at +0.6 V applied voltage are shown in Fig. 4.2.

There is no sense to present the comparison of experimental and theoretical data for the case of negative applied voltages; they coincide absolutely because during the model development, we have used time constants taken from experiments with negative applied voltages, when the reduction processes were supposed to take place simultaneously for all stripes on the whole channel length within the active zone. Therefore, the task was to make a model of kinetics for positive voltages and to explain its difference with respect to the case of negative voltages.

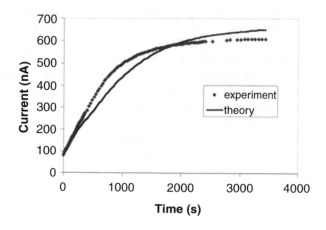

Fig. 4.2 Kinetics of the total current variation when +0.6 V was applied to the organic memristive device: experimental (rhombs) and theoretical (solid line) data. (Reprinted from Smerieri et al. [266], with the permission of AIP Publishing)

Fig. 4.3 Kinetics of the variation of current values in the circuits of the drain (brown line) and reference electrode (blue line). For convenience, the value of the current in the circuit of the reference electrode was multiplied by 10 and shifted by 0.21. (Reprinted from Smerieri et al. [266], with the permission of AIP Publishing)

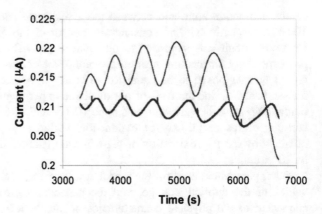

As it is clear from Fig. 4.2, the developed model explains rather well the experimental data. In particular, the model forecasts the 50% decrease of the total current with respect to its initial value during 100 seconds and the 90% decrease during 500 seconds for the applied −0.1 V and reaching 50% of the maximal value within 1700 seconds and 90% within 2700 seconds for the applied +0.6 V. It corresponds well to experimental results considered in Chap. 2.

The next step of the model application was connected to the explanation of the device function in the mode of generation of auto-oscillations connected to the presence of elements capable to accumulate charges in the reference electrode circuit. As it was explained above, it can be reached by connecting the external capacitor or by the use of graphite as the reference electrode material. For the model, it is not important which approach was used; it is important only the value of the effective capacity in this circuit.

Auto-oscillations were observed both for total and ionic currents with the phase shift, as is shown in Fig. 4.3.

Results of modeling of the system, containing 5 µF capacitor, when +5.0 V was applied, are shown in Fig. 4.4.

As it is clear from Fig. 4.4, the developed model allows to explain qualitatively the properties of the organic memristive device in the auto-oscillation mode. Despite the fact that the complete correspondence of experimental and theoretical results was not observed, the model described well several device characteristics observed in the experiment. First, the oscillations of the total current reach their maximum at about half a period later than the oscillations of the current in the circuit of the reference electrode. Second, the value of the oscillation period is not constant. Finally, the values of the oscillation amplitudes are also not constant.

Further testing of the model was done by the study of the influence of different variable parameters on the characteristics of the organic memristive devices when they work in the oscillation mode. Each parameter used within the model was varied and the range of its variation was often about two orders of magnitude. These steps implied the variation of only one parameter, all other parameters were fixed during

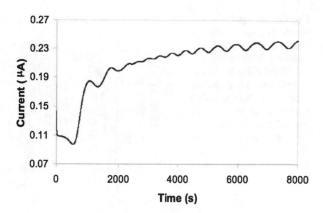

Fig. 4.4 Modelling results of the temporal behavior of the total current of the organic memristive device, containing 5 μF capacitor in the reference electrode circuit, when +5.0 was applied. (Reprinted from Smerieri et al. [266], with the permission of AIP Publishing)

these variations. After the calculations, the influence of the variable parameters on the characteristics of oscillations, such as amplitude and period of oscillations, as well as the final value of the total current, was analyzed. It was observed a significant difference in the influence of different parameters on the capability of the devices to generate auto-oscillations. This fact is in good agreement with experimental results: oscillations were observed only on some samples when a certain voltage was applied to the sample with a specific capacitor attached.

The only significant difference of the calculated results with respect to experimental ones was a trend of continuous increase of the total current value in time. This difference was attributed to the lack of the degradation parameter of the channel conductivity in the model. For removing this dismissing, a new parameter responsible for the gradual increase of the channel resistance in time was introduced into the model. The law, describing the temporal variation of the resistance, was taken from the experimental data of the device before the optimization of its architecture and composition [47, 194]. Other parameters of the model were not changed. Temporal behaviors of drain and reference electrode currents for such a modified model are presented in Fig. 4.5.

Summarizing the results of this section, it is possible to conclude that the developed model describes rather well the properties of the organic memristive device. In particular, this model was built for explaining two experimentally observed effects: difference in the resistance switching kinetics for positive and negative applied voltages and working of the device in the auto-oscillation mode when the circuit of the reference electrode contains an element capable to charge accumulation. The developed model was based on the phenomenon of the resistance variation of polyaniline channel, taking into account all currents in the active zone, composed by the contact of conducting polymer and solid electrolyte, as well as the variation of the distribution of the applied voltage on the channel length depending on the duration of the exposure of the particular segment of polyaniline channel of the device in the active zone to certain values of the potential (higher than the oxidation one or lower than the reduction one).

Fig. 4.5 Temporal
behavior of currents in the
drain (a) and reference
electrode (b) circuits,
calculated using the
modified model, taking into
account the degradation of
the electrical properties of
the active zone of the
organic memristive device.
(Reprinted Smerieri et al.
[266], with the permission
of AIP Publishing)

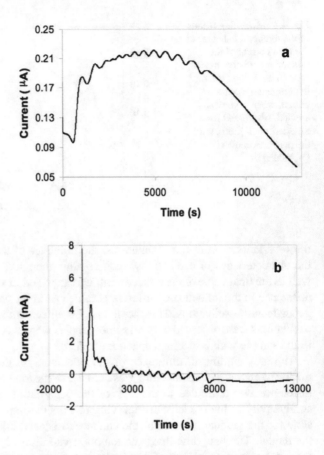

The developed model describes the difference in kinetics of the resistance switching during the application of positive and negative potentials with good quantitative correspondence to the experimental data. In the case of working in the auto-oscillation mode, good qualitative correspondence to experimental data was also observed.

4.2 Simplified Model of the Organic Memristive Device Function

As it has been shown in the previous section, the developed model describes rather well the properties of the organic memristive device. It has explained numerically also intuitively made hypothesis done at the initial stages of working with these devices responsible for the different device characteristics measured when they were biased with negative and positive voltages. However, the developed model is rather complicated and it requires significant calculation time even in the case of a single memristive device. The calculation of circuits, composed of several devices of such type, was practically impossible. Therefore, it was necessary to develop a simplified version of the model, allowing the calculation of features of systems composed of several devices [233].

Any memristive system, by the definition, can be described by dependences of the current $i(t)$ and voltage $v(t)$ on an internal parameter, dependent on time, as well as on current and voltage values. Consequently, it is possible to consider memristive devices, whose properties are determined by current or voltage, according to Eqs. 4.1 and 4.2.

$$i(t) = G(x, v, t)v(t) \tag{4.1}$$

$$\dot{x} = f(x, v, t). \tag{4.2}$$

The equivalent circuit of the organic memristive device that will be used for the development of the simplified model [233], is shown in Fig. 4.6.

The layer of polyaniline in the active zone is represented in Fig. 4.6 by the variable resistor because it can change its resistance according to the oxidation degree of the polyaniline film (fraction of oxidized PANI with respect to the whole amount). The area of the contact "conducting polymer – solid electrolyte" is represented by the capacitor, as this contact allows also charge accumulation.

As the resistance of the device depends on the oxidation degree (percentage of oxidized PANI molecules with respect to the total amount), and hence on the charge q in the polyaniline chain, we can express it by Eq. 4.3:

$$R(q) = R_{\text{off}} + q/q_{\text{max}} \cdot R_{\text{ox}} + (1 - q/q_{\text{max}})R_{\text{red}} \tag{4.3}$$

where q_{max} is the total number of segments, allowing oxidation, multiplied by the ion charge. R_{ox} and R_{red} are resistances corresponding to the completely oxidized or completely reduced states; respectively. R_{off} is the residual resistance of the polyaniline layer outside the active zone. In our case, q corresponds to the number of Li^+ ions moving between polyaniline and electrolyte. Thus, the derivative of this

Fig. 4.6 Equivalent circuit used for the development of a simplified model of the organic memristive device. (Reprinted from Springer Nature, Pincella et al. [233], Copyright (2011))

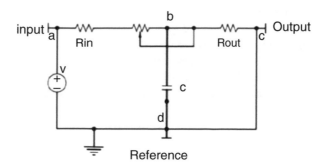

value is the ionic current. Charging/discharging of the capacitor in the equivalent circuit (Fig. 4.6) must occur through the equivalent resistor (Eq. 4.4):

$$R_{eq} = \frac{R_{in} \cdot R_{out}}{R_{in} + R_{out}} \tag{4.4}$$

It contains two resistors: R_{in} and R_{out}. Therefore, the equivalent voltage can be expressed according to Eq. 4.5:

$$V_{eq} = \frac{R_{out}}{R_{in} + R_{out}} \cdot V \tag{4.5}$$

The ionic current, responsible for the capacitor charging, can be expressed by Eq. 4.6:

$$\dot{q} = \frac{V_{eq} - \frac{q}{C}}{R_{eq}(q)} = \frac{\Delta V(q)}{R_{eq}(q)} \tag{4.6}$$

where $\Delta V(q)$ is the voltage between the active zone and the capacitor, resulting from the capacitor charging. In order to have a complete view, the previous equation must be accomplished by two probability factors (Eq. 4.7):

$$\dot{q} = \frac{\Delta V(q)}{R_{eq}(q)} \cdot \ n(q) \cdot \ P(\Delta V, V_{ox}, V_{red}) \tag{4.7}$$

where $n(q)$ takes into account the number of available states for redox reactions, while $P(\Delta V, V_{ox}, V_{red})$ takes into account the fact that it is necessary to be higher than V_{ox} or lower than V_{red} for starting the active component of the ionic current. Both mentioned factors are necessary for the nonlinear behavior of the device. In particular, $n(q)$ determines the saturation of the conductivity, while $P(\Delta V, V_{ox}, V_{red})$ takes into account the necessity of the activation potentials. Considering the current through the device (Eq. 4.8):

$$i(t) = V(t)/R(q) \tag{4.8}$$

and taking into account Eq. 4.7, we arrive at Eqs. 4.1 and 4.2, describing in general memristive devices. The number of oxidized states (q) is a system internal variable parameter in this case.

It is necessary to note that the value of the ionic current is usually about two orders of magnitude lower than the current in the polyaniline channel and also that the reference electrode is connected to the source, which allows considering the device as a two-terminal element with respect to the external circuit. Thus, the ionic component of the current can be written as in Eq. 4.9:

$$\dot{q} = \alpha(t) \cdot i(t) \tag{4.9}$$

and the resistance, consequently, can be expressed Eq. 4.10:

$$R(q) = R\left(\int_{-\infty}^{t} \alpha(\tau) \cdot i(\tau)d\tau\right) \tag{4.10}$$

Initially, the validity of this simplified model was checked for the explanation of the variation of device characteristics, when the electrolyte was used in liquid or gel form. Then the model was used for the description of characteristics of systems, containing several memristive devices.

4.3   Electrochemical Model

The scheme of the organic memristive devices adapted for the development of this model is shown in Fig. 4.7 [267].

The device is connected to the external circuit through two terminals: source (S) and drain (D), respectively. For clarity, we suppose that the S electrode is grounded, while the external voltage is applied to the D electrode (as the device architecture is rather symmetric, the choice of the grounded and potentiated electrodes is rather arbitrary). The range of the externally applied voltages will be $-1.0 - +1.2$ V. The electronic conductivity of polyaniline in the active zone with the length L depends on the ionic current that has passed through it. The parameter can be connected to the integral of the ionic current in the circuit of the reference electrode determined by redox reactions, occurring in the active zone. When ionic current passes through the electrolyte between polyaniline layer and silver wire, polyaniline can act as an anode, providing the oxidation reaction and increasing the channel conductivity or as cathode providing reduction reaction and, therefore, inhibiting the conductivity of the channel. The redox reactions can be described by Eqs. 4.11 [267]:

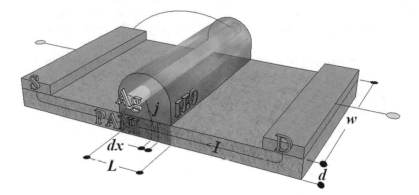

Fig. 4.7 Schematic representation of the organic memristive device in terms of the model of this section (not in the scale) [267]. (Reprinted from Demin et al. [267])

$$PANI^+Cl^- \text{(emeraldine salt)} + Li^+$$
$$+e^- \rightleftarrows PANI\text{(leucoemeraldine)} + LiCl,$$

$$Ag + ClO_4^- \rightleftarrows AgClO_4 + e^-. \tag{4.11}$$

It was assumed that the resistance switching can occur only in the area of polyaniline that is in a contact with a solid electrolyte. Oxidation and reduction of polyaniline take place at about +0.3 V and +0.1 V, respectively, with respect to the silver wire electrode that acts as a counter and reference electrode simultaneously [216].

From a physical point of view, oxidation and reduction processes can be expressed in terms of activation energies (or energy barriers) and their values can be expresses by Eq. 4.12:

$$E_{ob} = eV_{ob}$$
$$E_{rb} = eV_{rb} \tag{4.12}$$

where e is an elementary charge.

When the voltage V^* is applied to the polyaniline/electrolyte junction, the energy of the polyaniline oxidized state is shifted by $-eV^*$, due to the presence of additional charge. According to [267], the reaction rates can be expressed by Eq. 4.13:

$$\nu_{ox} = k_{PANI} \cdot n_{NH} \cdot n_{Cl} \exp\left(-\frac{eV_{ob}}{k_B T}\right) \exp\left(\frac{(1-\alpha)eV^*}{k_B T}\right),$$
$$\nu_{red} = k_{PANI} \cdot n_{NHCl} \cdot n_{Li} \exp\left(-\frac{eV_{rb}}{k_B T}\right) \exp\left(\frac{\alpha eV^*}{k_B T}\right), \tag{4.13}$$

where n_{NH} and n_{NHCl} are the volume concentrations of the reduced (amino groups NH-) and oxidized (HNCl-groups) sites in the polyaniline film, n_{Cl} and n_{Li} are the volume concentrations of Cl^- and Li^+ ions, k_{PANI} is a redox rate constant of polyaniline, α is so-called transfer coefficient of the reaction that can be in the range between 0 and 1, k_B is Boltzmann constant, and T is temperature. In further consideration, n_{NHCl} will be defined as "p," and p_{max} is the maximum of oxidized sites.

All amino groups of polyanilines without chlorine atoms are available for oxidation. Max possible number of such groups in the volume unit is $2p_{max}$. Thus, $n_{NH} = 2p_{max} - p$. Using the Boltzmann factor ($\beta = e/(k_B T)$), the total redox reaction rate can be expressed as Eq. 4.14:

$$\nu_{redox} \equiv \dot{P}$$

$$= k_{PANI} \cdot (2p_{max} - p) n_{Cl} e^{\beta((1-\alpha)V^* - V_{ob})} - k_{PANI} \cdot P$$

$$\cdot n_{Li} e^{-\beta(\alpha V^* - V_{rb})}. \tag{4.14}$$

In further considerations, we will not take into account the voltage difference at the silver/polyethylene oxide interface, as it is negligible with respect to that at the polyaniline/polyethylene oxide interface [268]. It seems a correct approximation because silver electrodes are widely used as reference electrodes due to fast chemical reaction kinetics [269].

Let us concentrate our attention on the redox reactions occurring in the polyaniline layer. Solid electrolyte, based on LiClO$_4$-doped polyethylene oxide, contains an excess of lithium ions. Cl$^-$ ions are present in the polyaniline layer due to its initial doping with HCl. Polyaniline is in a mostly conducting state called "emeraldine salt" when half of all amino groups are oxidized [216]. In our case, this form of polyaniline is used, and this is the reason for the introduction of factor 2 before p_{max}: the total quantity of all amino groups is the doubled max number of oxidized sites.

During the device functioning, lithium ions are involved also in the charge accumulation at the polyaniline/electrolyte junction capacitor, the lower plate of which, formed by polyaniline, is charged oppositely to the volume charge of the electrolyte. The capacitance of the polyaniline/polyethylene oxide junction will be considered as a constant along the whole active zone. This capacitor is not an ideal one: leakage ionic currents, due to redox reactions, can discharge the capacitor. Due to the presence of this active capacitor, our device can be considered as an extended memristive device according to the terminology introduced in [138].

When some segment of the polyaniline channel in the active zone is at a positive potential V^* higher than the equilibrium redox potential of the material, with respect to the silver wire, there is an electrostatic repulsion of Li$^+$ ions into the negatively charged electrolyte (determined mainly by ClO$_4^-$ ions). On the other hand, chlorine ions react with non-oxidized amino groups of polyaniline molecules, which results in the diminishing of charge, accumulated on the capacitor for the elementary unit e.

When the potential V^* is lower than the reduction one, lithium ions can enter into the polyaniline layer and participate in reduction reaction by electrostatic coupling with chlorine ions. In this case, the necessary electron is taken from the external circuit.

When all electrodes are disconnected, the charge at the polyaniline electrode is defined by open-circuit voltage V_{red}. When the silver electrode will be connected to one of the electrodes attached to the polyaniline layer, the device will work in a galvanic cell mode; the capacitor will be discharged till polyaniline will be completely reduced in the active zone, which will switch the device into the high resistance state. It is to note that freshly prepared devices are always in an insulating state before the application of any voltage.

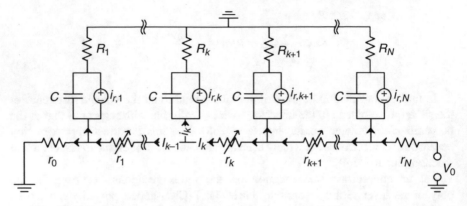

Fig. 4.8 Equivalent circuit of organic memristive device used for the model, developed in this section. (Reprinted from Demin et al. [267])

In order to take into account the nonideal nature of the capacitor, it is necessary to include other elements, as is shown in Fig. 4.8.

It is important that the leakage current in this circuit is represented by a current generator and not by a passive resistor. This is due to the fact that the value of the ionic current does not depend linearly on the voltage on the capacitor and can be expressed by Eq. 4.15:

$$I_{\text{redox}} = e \int_{0}^{d} \nu_{\text{redox}}(y) dy, \qquad (4.15)$$

where the integral is taken along the direction perpendicular to the polyaniline plane through the total thickness of this layer (Fig. 4.7).

It is not known a priori the distribution of the reaction rates on the thickness of the polyaniline layer. Thus, it was suggested that the mechanism is a homogeneous one. When a segment in the active zone has a potential lower than the reduction one, the ionic current is mainly determined by the lithium ions penetration into the polyaniline layer filling rather homogeneously the whole thickness of the layer, which results in the polyaniline reduction.

The suggestion on the homogeneous nature of redox reaction through the whole thickness is supported by its thin nature (usually, the thickness of the polyaniline channel is in the 20–100 nm range). Therefore, the density of the ionic current can be expressed as 4.16:

$$I_{\text{redox}} = e\dot{p}d. \qquad (4.16)$$

In order to have the possibility of the reaction rate estimation, it is necessary to evaluate values of concentrations n_{Cl} and n_{Li}. The density of bounded chlorine ions is determined by the number of oxidized sites p. During the reduction process, chlorine ions are detached from polyaniline and form an electrostatic complex with lithium ions. It results in the increase of the concentration of free chlorine ions (eq. 4.17):

$$n_{Cl} = p_{max} - p \tag{4.17}$$

In the case of the lithium ion concentration, according to the assumption about the homogeneity of the reaction on the whole thickness of the polyaniline channel, it can be written as (eq. 4.18):

$$n_{Li} = n_{Li,0} \tag{4.18}$$

where $n_{Li,0}$ is the known concentration of lithium ions in polyethylene oxide. The validity of this assumption will be tested, analyzing the simulation results.

According to the equivalent circuit (Fig. 4.8), the device is composed of a large number of circuit elements connected in series and in parallel. Similar to the first model, each of these elements can be attributed to a certain stripe of polyaniline in the active zone corresponding to the segment dx (L/N) in Fig. 4.7. Each of these elements contains a capacitor C with the charge q_k and resistor R_k, corresponding to the resistance of the electrolyte. The stripe resistance r_k is a variable one according to the ionic charge passed in this area due to redox reactions. Constant resistors r_0 and r_N correspond to the polyaniline areas outside the active zone. The voltage V_0, applied to the device is distributed along the channel length according to the actual values of variable resistors r_k. Therefore, the rate of redox reactions is not constant along the polyaniline channel.

Kirchhoff's equations for the kth segment can be written as 4.19:

$$\begin{aligned} i_k &= \dot{q}_k + i_{redox,k}, \\ V_k &= i_k R_k + \frac{q_k}{C}. \end{aligned} \tag{4.19}$$

These equations can be combined and give Eqs. 4.20:

$$\begin{aligned} \dot{q}_k &= \frac{V_k}{R_k} - \frac{q_k}{R_k C} - i_{redox,k}, \\ I_k &= I_{k-1} + i_k, \\ V_{k+1} - V_k &= I_k r_k, \end{aligned} \tag{4.20}$$

where V_k is a potential of the stripe. Other values can be expressed through the length of the stripe $\Delta x = L/N$ (eq. 4.21):

$$
\begin{aligned}
q_k &= \gamma(x)\Delta x, \\
R_k &= \eta(x)/\Delta x, \\
r_k &= \rho(x)\Delta x, \\
C &= \zeta \Delta x, \\
i_{redox,k} &= j_{redox}(x)\Delta x, \\
i_k &= j(x)\Delta x.
\end{aligned}
\tag{4.21}
$$

Linear and redox-related densities of the ionic current are j and j_{redox}, respectively (A/m units). The surface density of the current i_{redox} (A/m^2 units) on the segment with the width w will be (eq. 4.22):

$$
j_{redox} = w\, i_{redox}
\tag{4.22}
$$

Considering the above equations for the limit case (the number of stripes tends to infinity), we can obtain the following equations (prime corresponds to $\delta/\delta x$, dot corresponds to $\delta/\delta t$) (eq. 4.23):

$$
\begin{aligned}
\dot{\gamma} &= \frac{V}{\eta} - \frac{\gamma}{\eta\zeta} - j_{redox}, \\
I' &= j = \dot{\gamma} + j_{redox}, \\
V' &= I\rho, \\
j_{redox} &= e\dot{p}wd, \\
\dot{P} &= k_{PANI}\, e^{-\beta V_{ob}}(2p_{max} - p)(p_{max} - p) \\
&\quad \times e^{(1-\alpha)\beta(\gamma/\zeta)} - k_{PANI}\, e^{-\beta V_{rb}} p\, n_{Li,0} e^{-\alpha\beta(\gamma/\zeta)}.
\end{aligned}
\tag{4.23}
$$

The nonlinearity of the system is determined by the nonlinear rate of redox reactions (and, therefore, j_{redox}) with respect to charge density $\gamma(x,t)$. In the absence of j_{redox}, the system would be a linear one. The system will be a nonlinear one if the applied voltage V^* is higher than V_{ox} or lower than V_{red}.

The above equations can be simplified in the following way (eq. 4.24):

$$
\begin{aligned}
I' &= \frac{V}{\eta} - \frac{\gamma}{\eta\zeta}, \\
V'' - V'\frac{\rho'}{\rho} - V\frac{\rho}{\eta} &= -\frac{\gamma\rho}{\zeta\eta}, \\
\dot{\gamma} &= \frac{V}{\eta} - \frac{\gamma}{\eta\zeta} - e\dot{p}wd, \\
\dot{p} &= \widetilde{k}_{PANI}\, e^{-\beta(V_{ob}-V_{rb})}(2p_{max} - p)(p_{max} - p)e^{(1-\alpha)\beta(\gamma/\zeta)} \\
&\quad -\widetilde{k}_{PANI}\, p\, n_{Li,0}\, e^{-\alpha\beta(\gamma/\zeta)}.
\end{aligned}
$$

$$(4.24)$$

Thus, we have only three variables: p, γ, and V.

It is possible to add initial and boundary conditions (eq. 4.25):

$$
\begin{aligned}
p(x,0) &= P_0(x), \\
\gamma(x,0) &= \gamma_0(x), \\
V(0,t) &= 0, \\
V(l,t) &= V_0.
\end{aligned}
$$

$$(4.25)$$

It is to underline that during the analysis of the experimental results, the current decrease during the application of the voltage was fitted by two exponential dependencies [269]. If we consider that the capacitor charging time $\eta\zeta$ is significantly lower than the redox reaction time $1/\widetilde{k}_{redox}$ (where $\widetilde{k}_{redox} \sim k_{PANI}n_{cl}\exp\{\beta((1-\alpha)V^* - V_{ob})$ or $k_{PANI}n_{li}\exp\{-\beta(\alpha V^* + V_{rd})\})$, then the system 4.24 can be simplified during the time when the capacitor is relaxing to a quasi-equilibrium value, determined by 4.26:

$$\gamma \approx \zeta V - \eta\zeta e\dot{p}wd. \qquad (4.26)$$

This equation indicates that the variation of the potential distribution profile due to the changing of segment resistances takes place with characteristic time $1/\widetilde{k}_{redox}$. After this time, the value of the charge on the capacitor is determined by the component (eq. 4.26) responsible for a rather slow variation of ρ, I, V due to redox processes.

The presented system can be solved numerically with predetermined accuracy, taking into account initial and boundary conditions. Initially, $p(x,t)$, $\gamma(x,t)$, and $V(x,t)$ were calculated. After, it was possible to determine all other dynamic characteristics of the device.

The total current in the device can be divided into two components: electronic (I_e) and ionic, according to Eq. 4.27:

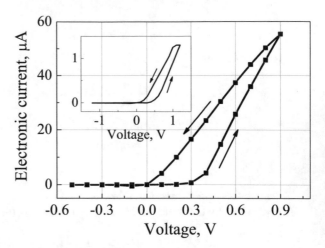

Fig. 4.9 Calculated voltage-current characteristics of the electronic component of the total current. Inset shows the experimental data. Arrows indicate the direction of the voltage variation; points correspond to the voltage values, where the calculations were performed. (Reprinted from Demin et al. [267])

$$I(x) = I_e + \int_0^x j(\xi)d\xi. \tag{4.27}$$

Inserting this expression to the previous ones, we have eq. 4.28:

$$V_0 = I_e \int_0^L \rho(x)dx + \int_0^L \rho(x) \int_0^x j(\xi)d\xi dx = I_e R + V_i, \tag{4.28}$$

where R is a total resistance of the polyaniline channel in the active area and V_i is a total potential difference of the whole polyaniline layer, due to the ionic current contribution. Thus, the electronic component of the current through the device can be expressed as Eq. 4.29:

$$I_e = \frac{V_0 - V_i}{R}. \tag{4.29}$$

The following parameters have been used for calculations: length of the polyaniline layer in the active zone $L = 1$ mm, width of the layer $w = 5$ mm, thickness of the layer $d = 20$ nm, $V_{ob} - V_{rb} = 0.3$ V, $\rho_{off} = 10^8$ Ω/cm, $p_{max} = 2.5$ 10^{21} cm^{-3}, $p_i = 10^{-3}$ p_{max}, $n_{Li,0} = 6.02$ 10^{19} cm^{-3}, $k_{PANI}n_{Li,0} = 0.47$ s^{-1}, $\alpha = 0.5$ [270], running capacity $\zeta = 200$ μF/cm, and electrolyte resistance $\eta = 2000$ Ω cm. All these parameters were taken directly from the experimental data or obtained by simple calculations.

Initial conditions corresponded to the insulating polyaniline layer in the active zone, experimentally observed situation for freshly prepared samples.

Calculated voltage-current characteristics for electronic and ionic currents are shown in Figs. 4.9 and 4.10, respectively. For the comparison, experimental characteristics are shown in insets. The time delay was 1 minutes for all points, as it was in the experimental conditions.

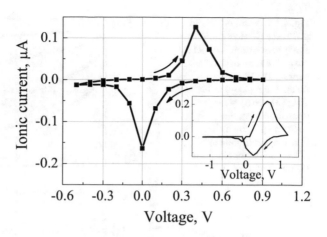

Fig. 4.10 Calculated voltage-current characteristics of the ionic component of the total current. Inset shows the experimental data. Arrows indicate the direction of the voltage variation; points correspond to the voltage values, where the calculations were performed. (Reprinted from Demin et al. [267])

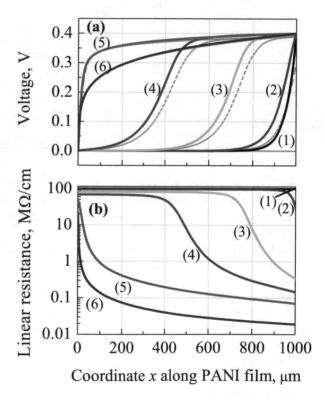

Fig. 4.11 Voltage (**a**) and resistance (**b**) distribution along polyaniline channel in the active zone at different moments of the device functioning: (1) –0 s; (2) – 0.5 s; (3) –3 s; (4) –6 s; (5) – 12 s; (6) –60 s (transition from the insulating to conducting states; applied voltage +0.4 V). (Reprinted from Demin et al. [267]).

The ON/OFF ratio for the calculated curves was three orders of magnitude, which is in good agreement with experimental data.

The developed model allows also to follow the variation of the potential distribution profile on the length of the polyaniline channel within the active zone. Such variation for the device transition from insulating to conducting forms for the applied voltage of +0.4 V is shown in Fig. 4.11a. Resistance distribution is shown in Fig. 4.11b.

Fig. 4.12 Voltage (**a**) and
resistance (**b**) distribution
along polyaniline channel in
the active zone at different
moments of the device
functioning: (1) –0 s; (2) –
12 s; (3) –20 s; (4) –60 s
(transition from the
conducting to insulating
states; applied voltage
−0.2 V). (Reprinted from
Demin et al. [267])

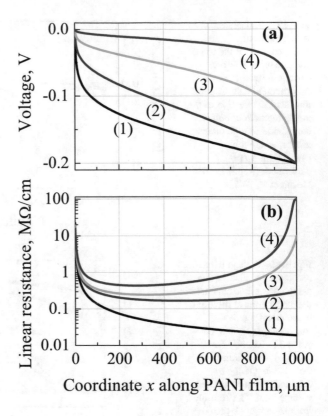

Coordinate x along PANI film, μm

As it has been discussed in the previous chapter, basing on intuitive considerations, the segments at higher potential transfer faster to the conducting state.

Dashed lines in Fig. 4.11 indicate the potential difference on the capacitor at the
polyaniline/electrolyte interface.

Similar characteristics, but for the case of the transition from conducting to
insulating states (applied voltage was −0.2 V), are shown in Fig. 4.12.

The developed model explains also what happens when the power supply is
switched off. In this case, we can suppose discharge of the capacitor and transfer of
polyaniline into the reduced state. It is in an agreement with simulation results that
are shown in Fig. 4.13 for the case when the applied voltage was switched from
+0.9 V to zero.

As it can be seen from Fig. 4.13, after the first rather fast process connected to the
capacitor discharge, we can observe a rather slow process of the polyaniline reduction. It means that the time necessary for the device transformation into its insulating
state must occur within a time of about 5 min, which differs from the experimental
data for about a factor of two.

To conclude, the proposed model is based on considering practically all processes
in the active zone of the organic memristive devices. It has demonstrated good
quantitative correspondence to experimental data connected to the voltage-current
characteristics and resistance switching kinetics. Therefore, it can be used for the

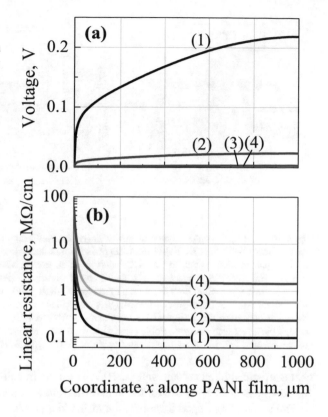

Fig. 4.13 Voltage (**a**) and resistance (**b**) distribution along polyaniline channel in the active zone at different moments after disconnecting the device from power supply: for (**a**) panel: (1) –0 s; (2) –0.1 s; (3) –1 s; (4) –60 s; for (**b**) panel: (1) –0 s; (2) –20 s; (3) –40 s; (4) –60 s. (Reprinted from Demin et al. [267])

further optimization of the device (materials, architecture) and the design of circuits based on these devices.

However, the last experimental data on the variation of the resistance profile on the length of polyaniline channel in the active zone [271] have demonstrated that even this model must be further developed. These results are discussed in the successive section.

4.4 Optical Monitoring of the Resistive States

In this case, we have used the property of polyaniline to change its color in accordance with its resistance state [272, 273]. The experiment was performed using the experimental set-up, shown in Fig. 4.14 [271]. The sample was illuminated with LED, and the image was acquired with an optical microscope, equipped with a CCD camera.

Mapping of the color distribution was done simultaneously with the electrical characterization (Fig. 4.14b). Optical measurements were performed at two

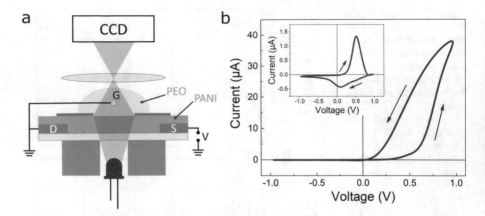

Fig. 4.14 (a) Experimental scheme of the spatially-resolved mapping of absorbance in a PANI-based memristive device: red light from LED goes through the active zone of a memristive device and is registered downstream by the CCD matrix of an optical microscope. The scheme of the assembly and electrical connections of a PANI-based memristive device: PANI film (green) on a glass substrate (light blue) with titanium Ti electrodes (dark grey) and an electrolyte layer (grey) with a silver Ag wire (light grey). The voltage V is applied to one of the substrate electrodes, two others are grounded. (**b**) A typical I-V curve of a PANI-based memristive device. The current through the silver gate electrode (ionic current) is shown in the inset. The arrows show the voltage sweep direction [271]. (Reproduced from Lapkin et al. [271], © 2020 Wiley-VCH GmbH)

characteristic applied voltage values: +0.7 V, for the transition from the insulating to conducting state and −0.2 V for the transition from conducting to an insulating state.

The conducting doped form of polyaniline is green in color, and it transfers into yellow after the reduction by the application of −0.2 V. It is possible to switch the device resistance back to the conducting state by the application the voltage of +0.7 V. Optical spectra of the active channel of the device in conducting and insulating states are shown in Fig. 4.15a.

As it is clear from Fig. 4.15a, a significant difference in the optical properties of the device in conducting and insulating states occurs in the wavelength range of 500–850 nm. Therefore, red AlGaAs LED with the center of the emission at about 660 nm was used. Variation of the absorbance at this wavelength was associated with polaron production in polyaniline connected directly to the material conductivity. During the experiment, micrographs of the sample (as it is shown in Fig. 4.15b) were acquired, and the absorbance was calculated for each spatial point along the applied electric field within the active zone of the device. The results, including a spatially resolved map of the optical absorbance, the evolution of the conductivity, and passed ionic charge during a cycle of conductivity potentiation-depression, are shown in Fig. 4.16.

As it is clear from Fig. 4.16, the application of −0.2 V results in the very fast transformation of polyaniline in the active zone to the reduced insulating state, however, not in the uniform manner. The reduction starts in segments that are closer to the silver wire, and then, it propagates in both directions to the source and drain

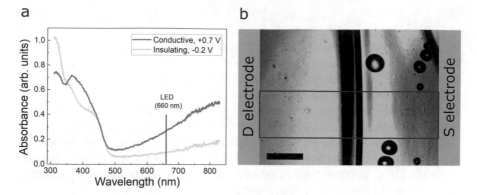

Fig. 4.15 Optical studies of a PANI memristive device: (**a**) UV-Vis spectra of a PANI film under an applied voltage of +0.7 (green) and − 0.2 V (yellow). The wavelength of the LED used to study changes in color is shown with the red line. (**b**) Optical microscopy image of the active zone used for spatially-resolved absorbance mapping. The area directly used for the extraction of absorbance values is shown with the red box. The black vertical line is the silver wire, the black circles are air bubbles in the electrolyte layer. The scale bar corresponds to 0.2 mm [271]. (Reproduced from Lapkin et al. [271], © 2020 Wiley-VCH GmbH)

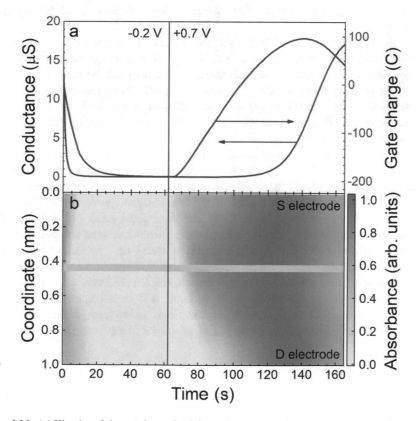

Fig. 4.16 (**a**) Kinetics of changes in conductivity and gate charge in time under applied voltages of −0.2 and + 0.7 V. The red vertical line indicates the voltage change. (**b**) Heatmap of absorbance evolution in time within the active zone under applied voltages of −0.2 and + 0.7 V. The grey horizontal line is the silver wire. The "source" electrode is on the top, the "drain" on the bottom of the panel [271]. (Reproduced from Lapkin et al. [271], © 2020 Wiley-VCH GmbH)

electrodes. However, we can see a significant increase of the resistance even before the moment when most of the polyaniline in the active zone is reduced. It is, obviously due to the fact that the total resistance of the channel in the active zone is mainly determined by the most insulating parts.

The situation is different when we apply +0.7 V for the induction of the transition from the insulating into the conducting state. The oxidation, in this case, starts for segments that are closer to the voltage application point. However, when the front of the potential distribution reaches the oxidation value in the central part (under the silver electrode), the rate of the transformation into the conducting state begins to be much higher for these central zone segments.

The faster kinetics of the reduction of polyaniline segments under the silver electrode at a voltage of -0.2 V can most likely be due to the limitation of the reaction rate by the current density of Li^+ ions in the electrolyte. Indeed, the distance between the Ag electrode and the PANI film is about an order of magnitude smaller than that between the silver wire and the "source" (or "drain") electrode ($\sim$40 μm vs. $\sim$400 μm). Accordingly, the current density (or, equivalently, the drift velocity) of Li^+ ions near the Ag electrode is $\sim$5–10 times higher than near the "source" (or "drain") electrode that specifies the reduction front propagation in PANI.

When the voltage of +0.7 V is applied, the conditions for oxidation reaction (overpotential is above +0.4 V) are met, first of all, near the point of the voltage application. In this case, the voltage profile along the active PANI channel redistributes in time (it falls mainly on the low-conducting yellow-color parts of the channel). Thus, when central segments arrive at the oxidation potentials, the ionic flow in these areas is more intensive because of the increased value of the electric field (geometrical considerations in the previous paragraph).

In this section, it has been demonstrated experimentally that electrochemical reactions responsible for the resistance switching in organic memristive devices occur nonuniformly within the active zone. Reduction and oxidation fronts propagate between the electrodes with some speed limited by the potential distribution in the active zone of the PANI layer, from one side, and by Li^+ ion drift velocity depending on the distance from the silver counter electrode, from another side.

The developed method of optical monitoring allows the nondestructive, intact, and continuous measuring of the resistive states of PANI-based memristive devices in various electronic circuits such as physical implementations of sensors with memory or artificial neural networks. In addition, it is an indication that even the last presented model must involve some corrections, taking into account the experimental data on the redox front propagation, presented in this section.

Chapter 5
Logic Elements and Neuron Networks

In this chapter, we will consider a hardware realization of logic elements with memory and simple artificial neural networks (perceptrons) based on memristive devices with special attention to organic memristive devices.

In general, living beings do not follow the laws of Boolean logic. Such operations, as classification and decision-making, depend not only on the actual configuration of the input stimuli but also on the experience accumulated in the past by the living being. Therefore, neuromorphic logic elements must include also the memorizing function. It must result in the realization of architectures, where, similarly to the case of the nervous system, the same functional elements will be used for both processing and memorizing the information. Such architectures will require the use of electronic elements with synapse-like properties.

As an example, we can consider the logical element "AND." For living beings, this function can be considered as the identification of the object when two or more essential properties are present. The example of such identification is shown in Fig. 5.1.

Two properties, shape and color, are used as input parameters, while the output corresponds to the identified object. Thus, in the case of Boolean logic, the presence of the spherical shape and the orange color must always result in the object identification (the orange). In the case of living beings, instead, this input information must be superimposed on the previous experience. In other words, as more frequently this configuration of input signals will be valid (confirmed by the feedback system, based on the taste sensations), the association of these properties with the object will be increased. Instead, if the taste receptors will show that this object is not an orange the association will be inhibited.

© Springer Nature Switzerland AG 2022
V. Erokhin, *Fundamentals of Organic Neuromorphic Systems*,
https://doi.org/10.1007/978-3-030-79492-7_5

Fig. 5.1 Illustration of the logic element "AND" functioning in living beings (identification of the object when two properties are present)

5.1   Logic Elements with Memory

In this section, we will consider the logic elements with memory, realized on memristive devices. On the one hand, these elements must perform main logic functions, such as "AND," "OR," and "NOT." On the other hand, the output signal of these elements must depend not only on the configuration of input signals but also on the previous history of the duration and frequency of their application [88, 274, 275]. Moreover, the value of the output signal in such systems must be constant in the absence of reinforcing or inhibiting signals on inputs. Thus, such elements will combine properties of traditional computer logic with synaptic memory, mimicking, to some extent the function of the brain, when the decision-making is based not only on the actual configuration of available stimuli but also on the experience, accumulated in the past.

All systems that will be considered in this chapter are based on the property of organic memristive devices to vary continuously their resistance under the application of constant voltages of certain values. This property was considered in detail in Chaps. 2 and 3. We only mention here that the output signal value variation can be described by Eq. 5.1 corresponding well to the experimental data [269].

$$I = A_1 e^{-\frac{t}{T_1}} + A_2 e^{-\frac{t}{T_2}} + C. \tag{5.1}$$

In this case, T_1 and T_2 are time constants, and C is a constant value, depending on the device resistance in saturation. As it was shown in Chaps. 2 and 3, this dependence is valid both for the increase and decrease of the conductivity. However, coefficients A_1 and A_2 will be different.

As it was considered in Chap. 3, the presence of two exponents is connected to two processes: charge/discharge of the capacitor and slow ionic motion due to redox reactions.

5.1.1   Element "OR" with Memory

The element OR is the simplest logic gate that can be realized with organic memristive devices. Its scheme is shown in Fig. 5.2.

The scheme contains two independent inputs (voltage sources), connected to the D electrode of the memristive device. The value of current in the circuit of the S electrode is an output signal. The values of input voltages are chosen in such a way that the application of any input must increase the conductivity of the memristive device. Therefore, the presence of the signal on any input will transfer the device into a more conducting state. The conductivity value in this case will depend on the duration of the signal applied to any input. The experimentally measured dependence of the output signal value on the time of the application of a signal to any input is shown in Fig. 5.3.

The presence of the constant offset voltage was used for maintaining constantly the state of the element when no input signal is applied.

The element performs the logic function OR, but the output signal is not in a binary code but can have intermediate values between 0 and 1 depending on the duration of the input signals application according to Eq. 4.2.

$$S_{out}(t) = \frac{I_{out}(t_1 + t_2)}{I_{out}(\infty)}.$$ (5.2)

In this case, t_1 and t_2 indicate time intervals when the signal was applied to input 1 or input 2, respectively, and the value $I_{out(\infty)}$ corresponds to the current value in saturation.

The conductivity of the memristive device and, therefore, the output value of the whole logic element will be increased (from 0 to 1 after the normalization) as soon as any of the inputs will be activated. The reached state will be maintained till the moment when successive stimulus will be applied. This stimulus can increase or decrease the value of the output signal (the decrease can be reached by the application of negative input voltage).

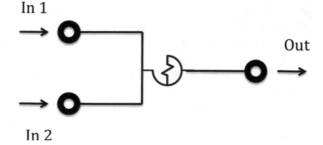

Fig. 5.2 Scheme of the logic gate OR with memory, based on the organic memristive device [274]. (Reproduced with permission from Erokhin et al. [274]. Copyright (2021) World Scientific Publishing Co., Inc.)

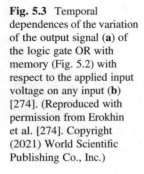 **Fig. 5.3** Temporal dependences of the variation of the output signal (**a**) of the logic gate OR with memory (Fig. 5.2) with respect to the applied input voltage on any input (**b**) [274]. (Reproduced with permission from Erokhin et al. [274]. Copyright (2021) World Scientific Publishing Co., Inc.)

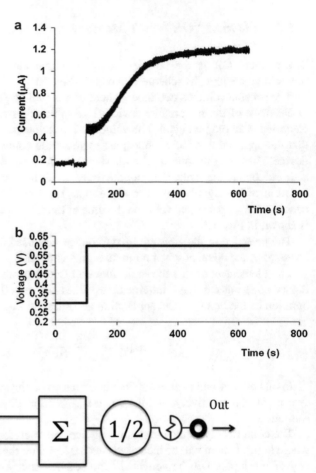

Fig. 5.4 Scheme of the AND gate with memory [274]. (Reproduced with permission from Erokhin et al. [274]. Copyright (2021) World Scientific Publishing Co., Inc.)

5.1.2 Element "AND" with Memory

The realization of this element is slightly more complicated with respect to the OR gate and requires the use of additional electronic components. The scheme of the AND gate with memory is shown in Fig. 5.4.

Two inputs of the circuit are connected to summator (operational amplifier, input voltages are passing through resistors with equal resistances) which provide the output voltage, equal to the sum of two input voltages. In order to have the same value of the output signal as it was in the case of OR gate (values of input signals are the same as in the case of OR gate), the circuit contains also a divider, allowing to have half of the sum of input voltages on the memristive element. The divider is based on the serial connection of resistors with equal resistance and readout of the

Fig. 5.5 Temporal dependence of the output signal of the element AND with memory (upper panel); dependences of the applied voltages on the input 1 (middle panel) and input 2 (lower panel) [274]. Reproduced with permission from V. Erokhin et al. Copyright (2021) World Scientific Publishing Co., Inc.

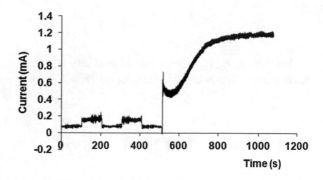

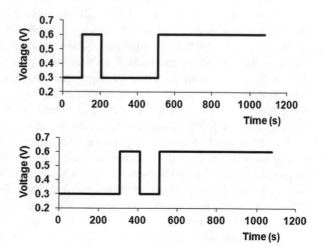

voltage from the point of their connection. In order to separate these resistors from inputs and output, they can be connected through operational amplifiers.

Temporal dependences of the output signal value on the activation of both inputs are shown in Fig. 5.5.

Individual application of signals to any input results in small and reversible increase of the output signal (according to Ohm's law). The situation is different when both input signals are applied simultaneously: we see the temporal increase of the output signal value. Thus, the function of this element is rather similar to associative synaptic learning. The reinforcement of the output signal demands in this case the presence of both input signals. In this case, we can see the increase of the connection of each input with the output, but this connection remains constant if both inputs are not activated simultaneously. The state of the output signal value of the gate AND with memory can be described by Eq. 5.3.

$$S_{out}(t) = \frac{I_{out}(t_{comm})}{I_{out}(\infty)}. \qquad (5.3)$$

In this case, t_{comm} is a duration of time interval when both inputs of the AND gate with memory are activated simultaneously.

5.1.3 Element "NOT" with Memory

The scheme of the gate NOT with memory is shown in Fig. 5.6.

This scheme contains two additional resistors, summator, and the external voltage generator, different from that used for the application of the input signal. The value of the resistor R_2 must be intermediate between the resistance of the memristive device in the conducting and insulating states. As typical values of the resistance of the memristive devices in the conducting and insulating states are 10 kΩ and 1000 MΩ, respectively, the value of R_2 was chosen to be 10 MΩ. The value of R_1 is not critical, but it must be less than R_2. The value of +V must be relatively small for avoiding the possibility of the memristive device transfer into the conducting state.

Before the application of the input signal, the whole external potential difference will be concentrated on the memristive device, as its resistance is two orders of magnitude higher than resistances of R_1 and R_2. It results in the high value of the output current. The application of the input signal, the value of which is comparable with those used in the case of AND and OR logic gates with memory, results in the memristive device switching to its conducting state, which will redistribute the voltage between the memristive device and resistors R_1 and R_2. The value of the output signal will be determined by Eq. 5.4.

Fig. 5.6 Scheme of the NOT gate with memory [274]. (Reproduced with permission from Erokhin et al. [274]. Copyright (2021) World Scientific Publishing Co., Inc.)

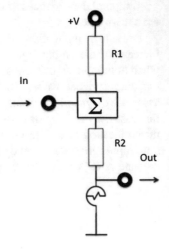

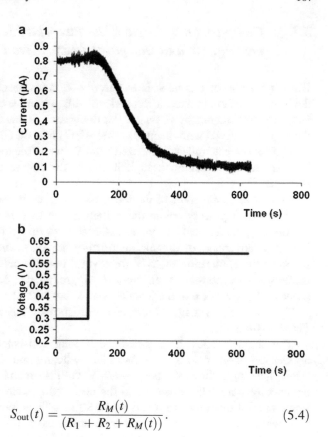

Fig. 5.7 Temporal dependence of the output signal of the NOT gate with memory (**a**); variation of the input signal (**b**) [274]. (Reproduced with permission from Erokhin et al. [274]. Copyright (2021) World Scientific Publishing Co., Inc.)

$$S_{out}(t) = \frac{R_M(t)}{(R_1 + R_2 + R_M(t))}. \tag{5.4}$$

$R_M(t)$ is the actual value of the resistance of the memristive device.

In the case of used values of the resistance of the resistors, this circuit allows to have about two orders of magnitude ratio of the output signal before and after the application of the input signal.

The temporal dependence of the output signal in this circuit is shown in Fig. 5.7.

Despite the fact that this scheme works as NOT logic gate with memory, it can be also considered as the analog of the inhibiting synapse. There are two reasons for this conclusion. First, the application of the input signal results in this case in the inhibition of the output one. Second, the inhibition degree depends on the duration of the input signal.

To conclude, all used gates with memory were considered for suggesting the possible realization one-bit full adder system [274].

5.1.4   Comparison of Logic Elements with Memory, Based on Organic and Inorganic Memristive Devices

Even if the aim of this book is to describe systems based on organic memristive devices, it is useful to make a comparison with properties of similar circuits realized with inorganic memristive devices. For this reason, it has been realized and studied the logic gate AND with memory, using Al_2O_3 thin film as an active layer [88].

The bottom electrode was realized from Pt, while the top electrodes were formed by evaporation of titanium through the mask, allowing to obtain regular arrays, as is shown in Fig. 5.8.

As it is usual for most of the inorganic systems, it was necessary to apply the electroforming process before the system began to exhibit memristive properties (by the way, the fact that it is not necessary to perform electroforming processes is an important advantage of organic memristive devices). The electroforming process involved the application of 15 V for several minutes between the top and bottom electrodes (the bottom Pt electrode was grounded). Before the electroforming process, the resistance of the junction was about 10^{10} Ω.

Typical cyclic voltage-current characteristics of this structure were shown in Fig. 1.3 (Chap. 1).

The characteristics demonstrate the bipolar resistance switching mechanism: application of about +7.5 V results in the switching into a highly conducting state, while the application of about -2.0 V results in the switching into the low conducting state. The resistance in the conducting state was found to be 250 Ω, while in the insulating one it was about 30 kΩ. Therefore, the resistance ON/OFF ratio was about two orders of magnitude in this case.

The characteristic shown in Fig. 1.3 allowed to design a logic gate AND with memory, similar to that, fabricated with an organic memristive device and shown in Fig. 5.4. The results obtained on such AND gate with memory based on Pt/Al_2O_3/Ti system are shown in Fig. 5.9.

Fig. 5.8 Sample, composed of Pt-Al2O3-Ti sandwich structure, used for the realization of AND gate with memory [88]. (Reproduced with permission from Baldi et al. [88]. Copyright (2014) IOP Publishing, Ltd.)

Fig. 5.9 Temporal
dependence of the output
current value (**a**) in logic
gate AND with memory,
based on inorganic
memristive devices. The
variations of the input
voltages, applied to input
electrodes are shown in (**b**)
and (**c**), respectively.
(Reproduced with
permission from G. Baldi,
S. Battistoni, G. Attolini,
M. Bosi, C. Collini,
S. Iannotta, L. Lorenzelli,
R. Mosca, J.S. Ponraj,
R. Verucchi, and
V. Erokhin, "Logic with
memory: AND gates made
of organic and inorganic
memristive devices,"
Semiconductor Sci.
Technol., 29, 104,009
(2014). Copyright (2014)
IOP Publishing, Ltd.)

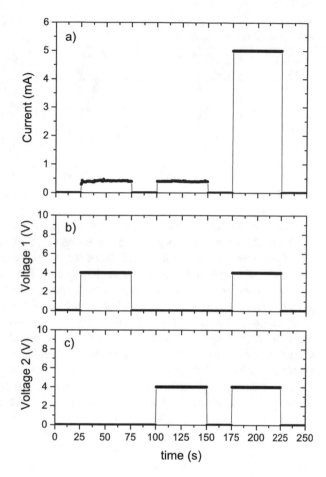

For the circuit shown in Figs. 5.8 and 5.9, the logical zero corresponded to the applied 0 V, while the logical one corresponded to the application of +4.0 V. Such choice of the value of "logic one" was determined by the fact that the device switching to the conducting state was possible only in the case of simultaneous application of both inputs (individual application of any input does not result in the variation of the resistance state of the device).

A comparison of the main characteristics of logic gates AND with memory realized using organic and inorganic memristive devices, is presented in Table 5.1.

As it is clear from Table 5.1, the properties of logic gates AND with memory, based on organic and inorganic memristive devices, are significantly different. The common feature is that the value of the output current was significantly increased only when both inputs were applied simultaneously and the increase of the output current was memorized. However, in the case of using inorganic devices, this transition was very fast: the saturation level was reached in less than 1 s. This conductivity level will be memorized and maintained even if only one input will

Table 5.1 Comparison of characteristics of logic gates AND with memory realized using organic and inorganic memristive devices

LOGICAL INPUTS	Output of inorganic element (mA)		Output of organic element (μA)	
	After 1 s	After 15 min	After 1 s	After 15 min
$In_1 = 0$ $In_2 = 0$	0	0	0	0
$In_1 = 1$ $In_2 = 0$	0.4	0.4	0.32	0.63
$In_1 = 0$ $In_2 = 1$	0.4	0.4	0.32	0.63
$In_1 = 1$ $In_2 = 1$	5.0 (compliance)	5.0 (compliance)	0.70	2.25

be applied in the future. Thus, it is possible to suppose that the area of application of such elements with memory, based on inorganic devices, will be connected to adaptive circuits, requiring the variation of the system properties, when two important stimuli are present. However, these elements will have a very limited application for the realization of neuromorphic systems, mimicking the properties of nervous systems. It is to note that the high switching rate, a very positive property in the case of traditional electronic devices, plays rather a negative role for neuromorphic applications because the kinetics of switching is weakly controllable. Thus, the circuit, based on inorganic devices, can be considered as a digital system with hardly resolvable intermediate states.

The situation is different in the case of organic memristive devices. The value of the output signal depends not only on the fact of the simultaneous presence of signals on both inputs but also on the duration (and/or frequency) of their simultaneous action. Such behavior is more similar to what happens in the nervous systems of living beings. At first glance, the work of such system is rather slow. However, considering that the ON/OFF ratio for such devices can reach 10^5 and the noise level is rather low, it is possible to state that switching between two well-resolved conductivity states (if it is necessary to transfer to binary codes) will be rather fast (characteristic time will be in ns range). Moreover, as it was shown in Chap. 2, the switching time of organic memristive devices depends significantly of their sizes. Thus, arriving at the sub-micron range will allow increasing further the switching rate.

In summary, the results of this section have demonstrated the possibility of the hardware realization of systems, where memorizing and processing of the information is done with the same elements, making a basis for bio-mimicking information processing. The next section will be dedicated to the realization of perceptrons – artificial neural networks with the ability to classify objects after appropriate learning.

5.2   Perceptrons

Perceptrons are particular versions of the artificial neural networks, designed for the performing of object classification. Neural networks are composed of layers of nonlinear threshold nodes, providing further propagation of the signal only when the integral of income signals will be higher than a defined threshold level. Each node in the layer is connected to all nodes of previous and successive layers. Connections between nodes have synapse-like properties: the possibility to vary their weight functions according to the applied training procedure. Currently, most artificial neural networks are realized at the software level, because it is rather easy to implement both threshold nodes and connections with variable weight functions. In addition, software implementation allows to know the state of the weight function for all connections, which is very important for effective training of the system. However, software implementation of artificial neural networks has a significant limitation. Traditional computers can perform only one operation in time (the situation is slightly better in the case of multi-nuclear computers, where several operations, corresponding to the number of nuclei, can be performed simultaneously). Therefore, parallel information processing is impossible in such systems. For this reason, hardware implementation of artificial neural networks can provide a real breakthrough in neuromorphic information processing.

Schematic representation of artificial neural network (in particular, perceptron) is shown in Fig. 5.10.

Due to synapse-mimicking properties, memristive devices are widely considered in the literature to be used as weight function varying connections [276–280].

It is possible to say that the first memristive devices-based implementation of artificial neural networks was reported when a single layer perceptron was realized [79]. The perceptrons [2, 281] were originally designed for performing classification tasks after the application of adequate training procedures [281, 282]. Single-layer elementary perceptron based on the organic memristive devices [283] which are the subject of this book, was realized only a few months after the inorganic one.

The limitation of the single-layer elementary perceptron is the fact that it can classify only linearly separable objects. In order to classify linearly not separable

Fig. 5.10   Scheme of the artificial neural network

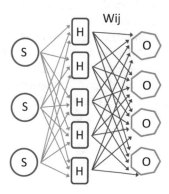

objects, we must have at least double-layer perceptrons. Hardware implementations of such systems were reported for organic [284, 285] and inorganic devices [286].

Research activities in the field of the use of memristive devices for artificial neural networks are widely presented in recent publications, indicating the importance of this application [287–295].

Before the detailed consideration of perceptrons, based on organic memristive devices, it is necessary to mention one other type of artificial neural network – reservoir computing system [296]. This system will allow the classification according to the variation of temporal sequences of the input signal. It can be used as a complementary counterpart to perceptron units that will allow to take into account also temporal variations of the object's properties. Up to now, few works on memristive devices were published on this subject [297]. However, organic materials seem to be good candidates for such systems. The reasons for this statement will be discussed after the consideration of networks with stochastic architecture in Chap. 6.

5.2.1 Single Layer Perceptron

The first realization of the artificial neural network, based on organic memristive devices, was a single layer elementary perceptron introduced by Wasserman, where each sensor neuron was directly connected to each neuron in the associative layer [281]. This type of perceptron does not require intermediate neurons and it contains only one layer of connections with variable weight functions.

Classification according to the NAND logic function was chosen as the task, because the objects, in this case, are linearly separable, as is shown in Fig. 5.11 [283].

The scheme of the realized perceptron is shown in Fig. 5.12.

The sensor layer contains two inputs, x_1 and x_2, corresponding to features of the object that must be classified. It contains also a permanent bias input x_0 that provides an offset, necessary for the separation of objects. Of course, such a simple system can distinguish only two classes of objects: if the output signal is one, the object corresponds to class A; if the value is zero, the object belongs to class B. The number of classes can be increased by adding neurons in the output layer and providing appropriate synaptic connections.

After the circuit was realized, it was necessary to identify the training algorithm. In this case, we decided to use the simplest method suggested by Rosenblatt – the method of error correction [282]. This algorithm implies the variation of all active connections in the same way during each iteration step according to the sign of the error between the actual and desirable values of the output signal, but not on its magnitude.

As it was mentioned above, this perceptron must perform the classification of objects according to the NAND logic function. Therefore, the values of input signals were attributed to "1" and "0." The output signal was also in binary codes when "1"

Fig. 5.11 Geometrical
representation of the linear
separability of objects,
corresponding to NAND
logic function
[283]. (Reprinted from
Demin et al. [283],
Copyright (2015), with
permission from Elsevier)

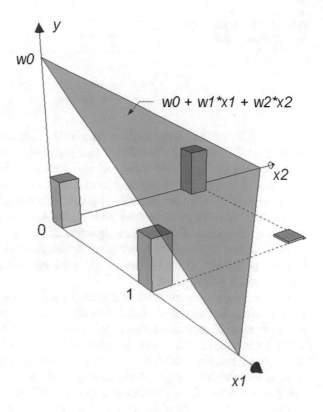

Fig. 5.12 Scheme of the
elementary single-layer
perceptron based on organic
memristive device.
(Reprinted from Demin
et al. [283], Copyright
(2015), with permission
from Elsevier)

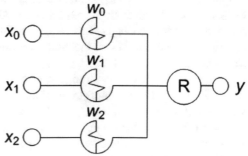

means that the object corresponds to the NAND class, while "0" means that it does
not. For example, as it is shown in Fig. 5.11 for the NAND function, vectors (0, 0),
(0, 1), and (1, 0) belong to the class "1," while the vector (1, 1) belongs to the class
"0." There is a first-order plane that can separate these objects.

Applied voltages were used as input signals and the current value was used as an
output. Before working with perceptron, it was necessary to identify input values
that will not vary the conductivity state of the memristive devices in the circuit.
Basing on the experimental and simulation information (Chaps. 2 and 4), +0.4 V was
chosen as logical "1" and +0.2 V as logical "0."

Table 5.2 Truth values for the NAND functions

In$_1$	In$_2$	Out
0	0	1
1	0	1
0	1	1
1	1	0

The training of the perceptron was done in the following way. First, it was built the truth table (Table 5.2) where theoretical values of the expected output signal are corresponding to all possible combinations of input signals.

During each epoch of the training procedure, all possible configurations of input signals (x_1 and x_2) were successively applied to the system (four possible combinations). After measuring the output signals, the sign of the error was calculated, comparing the value of the actual signal with the truth table. If this error was negative, it was necessary to apply potentiation stimuli to the memristive devices. If it was positive, it was necessary to apply the inhibition stimuli. If the error was zero, no action was required.

One training epoch consisted of successive application of all possible four input vectors, comparing the output value with the truth table and weight adaptation after the application of each input vector, when necessary. The training procedure was stopped when the error values were zero for all applied input vectors (in reality, it was repeated two times more in order to guarantee the success of the training).

In addition to the voltage values, corresponding to "1" and "0" inputs, two other voltages were used for the reinforcement (+0.7 V) and inhibition (−0.2 V) of the weight functions. The time intervals necessary for performing such weight function variations were taken from the experimental kinetics of the resistance switching of the used devices (see Chaps. 2 and 4). Therefore, these time intervals were 400 s for the reinforcement and 50 s for the inhibition (however, the duration of these intervals was varied).

In the case of this perceptron, the current value was considered as the output signal. In order to have a binary code of the output signal, two values of current were identified: threshold current (I_t) and demarcation current (I_d). The output was considered as "0" if the output current was less than I_t–I_d, and it was considered as "1" if the output current was more than $I_t + I_d$. The demarcation current was introduced in order to avoid misclassification of objects when the values of output current were about I_t. The values of I_t and I_d were 3.0 μA and 0.5 μA, respectively.

The results of training with different duration of potentiation and depression stages are shown in Fig. 5.13.

In the case of 600 s for the potentiation and 30 s for the depression, 15 steps were required for performing the classification according to the NAND function. As it is clear from Fig. 5.13, decreased time of the training stages requires an increased number of the training steps. Thus, there is an optimal duration of stages when the entire training procedure requires less time to be completed. In our case, this optimal potentiation time was found to be 200 s. Values of I_t and I_d also influence the required number of training steps. Increasing the values of these currents results in an

Fig. 5.13 Dependence of
the error sign during the
perceptron training to
perform NAND function on
the training step number for
the duration of potentiation/
depression stages: (1) 600/
30 s, (2) 200/20 s, (3) 150/
15 s [283]. (Reprinted from
Demin et al. [283],
Copyright (2015), with
permission from Elsevier)

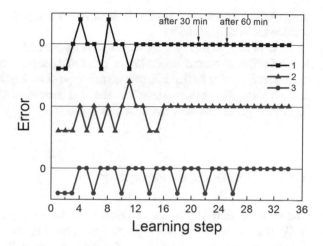

Table 5.3 Truth values for
the NOR functions

In$_1$	In$_2$	Out
0	0	1
1	0	0
0	1	0
1	1	0

increased number of required steps. After training, the realized configuration of
conductivities was maintained constant for at least several hours.

When the perceptron was trained to perform the classification according to the
NAND function, the attempt to re-train it for the classification according to the NOR
function was performed. This was done using the same parameters of all voltage and
current values. Of course, the other truth table (Table 5.3) was used.

In this case, only two steps with the duration of the potentiation stage of 200 s
were required. These steps were for the (0, 1) and (1, 0) combinations of the input
signals, because there is a difference of the output values between NAND and NOR
functions for these configurations only (compare Tables 5.2 and 5.3). After these
steps, the output error was zero for all input combinations.

As the training ability of the organic memristive devices-based perceptron has
been successfully demonstrated, the meaning of the schematic illustration of the
classification event shown in Fig. 5.11, can be understood better. Bars in this figure
correspond to the output values for each vector of inputs for the NAND function.
The plane is a first-order plane that can be described as $y(x) = w_0 + w_1 x_1 + w_2 x_2$, and
it must separate objects that correspond and not correspond to the NAND logic
function criterion. The conductivity of permanently biased memristive device w_0
defines an offset of this plane. Two other memristive weights determine the inclination of straight lines, intersecting the axis $y(x)$ with x_1 and x_2 coordinates. Training
of the perceptron continues until the plane will be shifted to a position, separating
selected classes of input signals. In this respect re-training from NAND to NOR

implies the increase of the plane inclination so that points (0, 1) and (1, 0) will be outside the classified objects.

In summary, in this section, we have demonstrated the possibility of the realization of artificial neural networks, in particular, single-layer perceptron, based on organic memristive devices. However, this type of perceptron allows to classify only linearly separable objects. Therefore, the next section will be dedicated to a double-layer perceptron that does not have this limitation.

5.2.2 Double Layer Perceptron

The logic continuation of the work on a single-layer elementary perceptron is the realization of a double-layer perceptron, allowing the classification of linearly non-separable objects. XOR logic function was used for training this perceptron, as the objects to be classified are not linearly separable. Referring to Fig. 5.11, two diagonal values, in this case, must be "1," while the other two values must be "0." Obviously, there is no first-order plane that can separate these classes of objects.

The scheme of the double-layer perceptron is shown in Fig. 5.14.

The scheme contains two inputs (X_1, X_2), two neurons in the hidden layer (the number can be increased) and one output neuron (again, the number can be increased) [284].

Even if the complication with respect to a single-layer perceptron seems to be not very significant at first glance, it required the realization of much more complicated circuits, as it will be shown below.

Inputs and neurons in the hidden layer are connected by links with synaptic weights (w_{ij}, w_{jk}). The circuit diagram of the network, based on organic memristive devices, is shown in Fig. 5.15. Areas, limited by circles, correspond to circles (neurons) in Fig. 5.14.

Each weight was represented by two memristive devices. The ability to vary the resistance (synaptic weight) of every memristive device independently from others is a critical requirement for training the network. An access system, based on CMOS transistors, providing voltage-controlled switching, was developed for this reason. This system was used for making two actions: reading the voltages during the information processing and application of voltages to selected memristive devices during the training procedure. A commutator composed of the 1-in-8 analog switch

Fig. 5.14 Scheme of the double-layer perceptron [284]. (Reproduced from Emelyanov et al [284] under permission of a Creative Commons license)

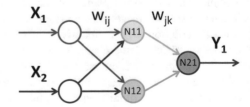

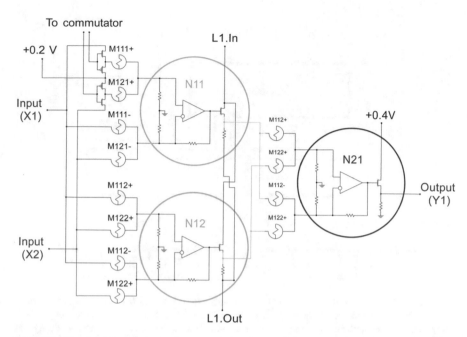

Fig. 5.15 Circuit diagram of organic memristive device-based hardware artificial neural network. Neurons are marked by circles and consist of differential summator and activation function. Numeration of memristive links (Mnij(+/−)) comprises a number of layer n, connected to *i*th input and *j*th output neurons. A sign defines if this partial weight is positive or negative. The access system is shown for M111+ and M121+ memristive devices and is omitted for others for simplicity [284]. (Reproduced from Emelyanov et al. [284] under permission of a Creative Commons license)

(considered as "master") and two others ("slave") connected in series allowed to control all 12 switches in the circuit by five logic inputs (Fig. 5.16).

The body of the artificial neuron (soma), providing two important functions: summation and threshold, was implemented by an op-amp-based differential adder and a voltage divider with a MOSFET controlled by the output of summator. The differential summator is required for the separation of different classes of input combinations according to Equation $_{5.5}$:

$$y = \sum w_i x_i \qquad (5.5)$$

where *y* is the output voltage of the summator and x_i and w_i are input voltages and corresponding weights, respectively.

The scheme allows also the realization of negative synaptic weights by doubling the number of memristive devices. This is critical for the training algorithms convergence in almost all possible tasks. Each synapse was represented by two memristive devices, "excitatory" and "inhibitory," connected to inputs of the op-amp. The resulting weight of *i*th synapse is (eq. 5.6):

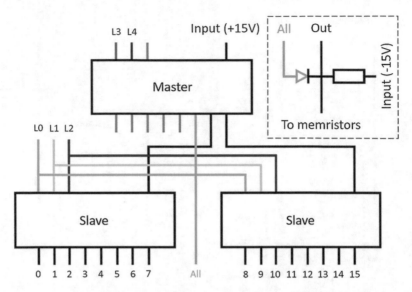

Fig. 5.16 Logic scheme of the commutator used with 5 logic inputs (L0-L4) and 16 outputs (only 12 of them were used according to the number of memristive devices). Separate output "All" corresponds to the application of control voltage (+15 V) to all memristive devices access system (during reading some input vector by the perceptron). In the absence of the control voltage, −15 V was applied to the access system due to the necessity of applying +0.2 V to all memristive devices [284]. (Reproduced from Emelyanov et al. [284] under permission of a Creative Commons license)

$$w_i = R_{\text{fb}}(G^+{}_i - G^-{}_i), \tag{5.6}$$

where R_{fb} is the value of the feedback resistance and $G^+{}_i$ and $G^-{}_i$ are conductances of the ith excitatory and inhibitory memristive devices, respectively.

The output voltage Y was applied to the gate of the field elect transistor in the voltage divider connecting the neuron output to logic "1" when it is open and to "0" when it is closed. The threshold voltage of the divider depends on the characteristics of the used MOSFET, and it was 1.8 V in our case. The typical transfer function (activation function) is shown in Fig. 5.17.

Similar to the case of single-layer perceptron, the voltage of +0.4 V was considered as a logic "1" and + 0.2 V as logic "0." Voltages of +0.6 V and − 0.2 V were used for the potentiation and depression, respectively.

Since a double-layer perceptron is able to solve linearly non-separable tasks, the classification according to the "XOR" logic function was chosen for the network training. This task cannot be solved by an elementary (single-layer) perceptron where each output neuron implements one hyper-plane, separating the classes. Nevertheless, the second layer neurons in a double-layer artificial neuronal network perform the separation in a space of the first layer, enabling union, intersection, and difference of the "subclasses" highlighted by the hidden layer of the network.

Fig. 5.17 Transfer
functions of the three used
voltage dividers
implementing an activation
function of neurons
[284]. (Reproduced from
Emelyanov et al. [284]
under permission of a
Creative Commons license)

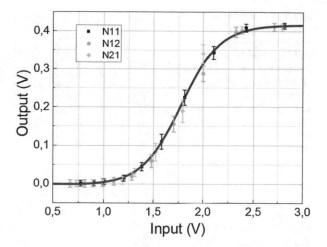

In machine learning, the back propagation with batch correction learning algorithm [298] is widely used for nonlinearly separable task solving. The algorithm
comprises the calculation of the gradient of a squared error function with respect to
all the weights in the network. The gradient is fed to the optimization method which
uses it to update the weights for minimizing the squared error function. It requires a
very precise variation of weight values. This item was a critical one for the hardware
perceptron realization due to the fact that resistive switching kinetics of memristive
devices were not similar enough for the unified mathematical model. Therefore, it is
possible to follow only the weight correction direction (sign) but not its value,
choosing the empirically established training pulse time duration. Such modification
of the back propagation learning algorithm leads to the strong correlation of the
necessary number of steps to converge with the initial weight distribution: the closer
it was to the final distribution, the less number of steps was required. It is to note that
even not all initial states of the network led to the convergence. Possible solutions to
this critical issue could be an implementation of other algorithms based on spike-
timing-dependent plasticity (STDP) rules [211, 212, 299] or realization of the circuit
where the conductivity of each element would be measured with a contactless
spectrophotometric method [271, 272].

Each step of the training procedure consisted consecutively of the application of
the whole set of vectors $x(k)$ ($k = 1, 2, 3, 4$), actual weight measuring (applying the
"reading" pulses), and weight correction (applying the "writing" pulses). The correction pulse duration values were chosen to minimize the duration of training steps,
and it was kept constant (but different for depressing and potentiating pulses) for all
steps in the whole learning procedure. The procedure was performed until the
convergence. Figure 5.18a shows the results of the learning procedure for the
XOR logic function at the first and the last iterations. Figure 5.18b depicts the
weight values change after learning. As described above, each weight was adjusted
by two memristive devices (their conductances are not shown separately) and set in

Fig. 5.18 Experimental data. (**a**) Output signal within the epochs before (left) and after (right) training and expected output signal (dotted). (**b**) Synaptic weights and (**c**) corresponding feature plane partition (area above and below the plane $y = 4.5$ is the class "1" and "0," correspondingly). Obtained separating planes are implemented by corresponding neurons in the first layer [284]. (Reproduced from Emelyanov et al. [284] under permission of a Creative Commons license)

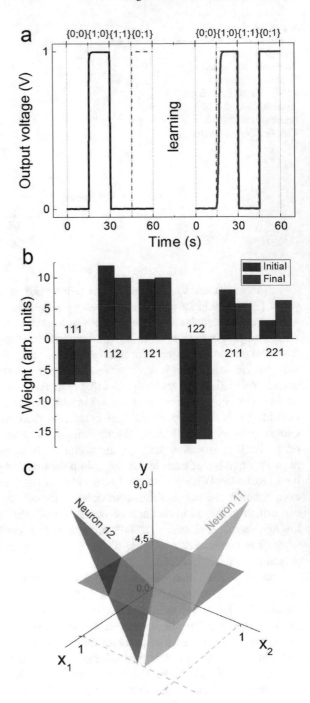

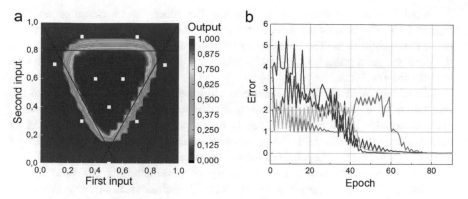

Fig. 5.19 (a) Simulated output signal and corresponding separating lines. (b) Error function values for training procedures, corresponding to four sets of initial weights [284]. (Reproduced from Emelyanov et al. [284] under permission of a Creative Commons license)

arbitrary units. As shown in Fig. 5.18c, the weights were adjusted so that two output classes were separated by two planes in the feature space.

The double-layer perceptron separates the feature space into different classes by the combination of hyper-planes. This feature permits to use this perceptron to classify objects not only in binary codes but also in the case when inputs can have intermediate values between "1" and "0." In other words, it can classify analog input signals. This possibility has been demonstrated by simulating the classification of a simple polygon – triangles. The model system contained two inputs, corresponding to planar coordinates of the points in a plane. As a straight line is performed by one neuron in the hidden layer, the first layer contained three neurons. The second (last) layer contained one neuron, and its output corresponded to whether the object is inside or outside the determined triangle.

The circuit was simulated using experimentally measured characteristics of real organic memristive devices.

The activation function of the ith neuron was obtained by fitting experimental data (Fig. 5.18) by sigmoidal function (eq. 5.7):

$$y_i = 1/\left(1 + \exp\left(\left(\textstyle\sum_i - 4.5\right)/0.5\right)\right) \qquad (5.7)$$

where Σ_i is the weighted sum of inputs of the ith neuron, considering +0.4 V as logic "1."

The training was performed by a simplified back propagation algorithm, replacing the derivative of the activation function by a constant 0.5 in order to speed up the convergence.

Possible positions of separating lines, implemented by hidden neurons, and the calculated output signal map are shown in Fig. 5.19a.

Vector points of the input data (white squares) were chosen for training the perceptron to classify analog signals in the triangle geometry. The points inside the triangle were classified as "1," while the others were classified as "0." Figure 5.19b shows how the square errors depend on the epoch number according to the initial conditions. The error convergence to zero indicates the applicability of a double-layer perceptron for the classification of objects in analog form for different sets of initial weights.

In summary, this section has demonstrated experimentally the possibility of the realization of a double-layer perceptron based on organic memristive devices, capable to make classification of linearly non-separable objects. Model, based on real characteristics of these devices, has demonstrated the principal applicability of this perceptron for the classification of analog signals. Further development of this study is planned to involve the possibility of non-disturbing real-time optical monitoring of the propagation of redox fronts in such devices that was described at the end of Chap. 4. It will allow not only to reveal the conductivity state of each weight in the network but also to follow the kinetics of their variation.

Chapter 6
Neuromorphic Systems

In this chapter, we will consider neuromorphic applications of organic memristive devices. The term "neuromorphic" is used here according to the criteria presented in the INTRODUCTION.

6.1 Learning of Circuits Based on a Single Memristive Device

6.1.1 DC Mode

The main feature of the neuromorphic system implies the possibility of its learning. In other words, the realized network must react to the variable external stimuli according to the executed training procedure (supervised learning) or basing on the accumulated experience (unsupervised learning) without any modification of the system architecture. The first step in this direction is the testing of the possibility of using a single memristive device, an analog of biological synapse in the nervous system, as a key element in such networks.

In order to demonstrate such possibility, the system with two input and one output electrodes has been chosen [300]. The system can be compared in a very simplified manner, as was done later in a lot of cases [155, 211, 242, 299, 301–305], with a Pavlov dog learning (classic conditioning). It was assumed that one input corresponds to the food presence, while the other one to the bell ring. Before training, the acoustic signal (neutral stimulus) did not result in the appearance of a significant output signal connected to the, for example, salivation. However, the conditioning took place during the simultaneous application of both stimuli (food and sound). After this conditioning (learning), the application of the initially neutral stimulus

© Springer Nature Switzerland AG 2022
V. Erokhin, *Fundamentals of Organic Neuromorphic Systems*,
https://doi.org/10.1007/978-3-030-79492-7_6

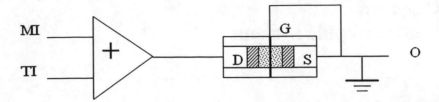

Fig. 6.1 Scheme of the experiment on the learning capability of the organic memristive device [300]. (Reprinted from Smerieri et al. [300], Copyright (2008), with permission from Elsevier)

resulted in the appearance of a significant output signal. Learning, in this case, was connected to the association of neutral signal (bell ring) with the presence of food.

The scheme implemented for the realization of such learning is shown in Fig. 6.1.

As was mentioned above, the circuit includes two inputs: MI (main input) corresponds to the acoustic stimulus, and TI (training input) corresponds to the presence of food. Similar to previous Chap. 5, dedicated to artificial neuron networks, voltages were used as input stimuli. Values of these voltages (+0.3 V) were chosen in such a way that been applied individually, the potential difference is not enough for switching the memristive device to a highly conducting state. Both inputs were applied to the summator, providing the output signal, equal to the sum of input voltages. The value of the current in the source-drain circuit of the memristive device was considered as the output. It was assumed that the value of the output signal must be higher than a certain threshold value before the execution of a defined function (salivation, in the case of Pavlov dog learning mimicking) could be possible.

Temporal dependence of the output current during the training of the circuit in Fig. 6.1 is shown in Fig. 6.2.

The experiment was started when both inputs were at zero voltage. After 250 seconds (point 1), only MI, corresponding to the acoustic stimulus, was applied to the circuit. Its value was chosen in such a way that it does not change the conductivity state of the memristive device. It has resulted in the appearance of a current of about 0.14 µA, and this value was constant in time. Point 2 (550 seconds after the beginning of the experiment) corresponds to the situation when both MI (acoustic stimulus) and TI (presence of food) were applied simultaneously. The learning process was started at this moment. We can clearly see the gradual increase of the output current from 0.32 µA to 0.5 µA. After coming to the saturation (point 3), only MI stimulus was applied again to the circuit. The value of the observed current was 0.23 µA, which corresponds to about 160% of the value observed in the system before training at the same value of the input voltage.

If we suppose that the threshold value required for the function execution (salivation) is 0.2 µA, the system will not react to the presence of acoustic stimulus only before training while after it, the same stimulus will be associated with the presence of food, triggering the salivation function.

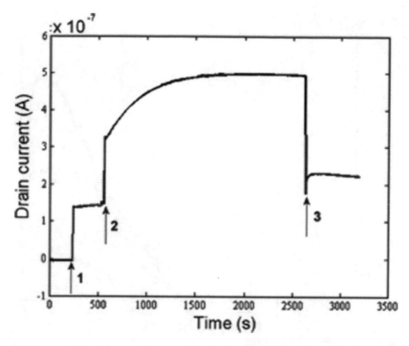

Fig. 6.2 Temporal dependence of the output current of the circuit, shown in Fig. 6.1. Arrows indicate the beginnings of the input voltages application: 1. MI – ON, TI – OFF; 2. MI – ON, TI – OFF; 3. MI- ON, TI – OFF. (Reprinted from Smerieri et al. [300], Copyright (2008), with permission from Elsevier)

Reported results indicate that the simplest training is possible even in circuits containing only one memristive device.

The more complicated learning demands, of course, the realization of more complicated systems containing a large number of memristive devices. In particular, an extremely high degree of the integration of synapse-like elements in restricted space can be achieved in systems, based on organic memristive devices with stochastic architecture that will be considered in successive sections. In these systems, we cannot be sure that source and reference electrodes will have the same potential, which is connected to the random distribution of materials in the network. Therefore, a special study was carried out for the investigation of the organic memristive device in the three-electrode configuration, applying different potentials to the reference electrode [300].

Cyclic dependences of the total and electronic currents of the devices at different voltages applied to the reference electrode are shown in Fig. 6.3.

As it is clear from Fig. 6.3, the resistance of the device increases significantly with the increase of the value of the positive potential, applied to the reference electrode. This behavior is connected to the fact that polyaniline in the active zone has more negative potential with respect to the reference electrode, which results in its

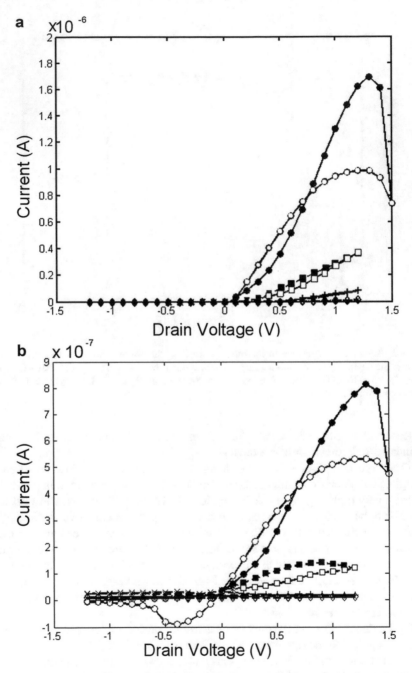

Fig. 6.3 Dependences of the total (**a**) and electronic (**b**) currents of the organic memristive device at different potentials, applied to the reference electrode: 0 V (circles); 0.2 V (squares); 0.4 V (pluses); 0.6 V (rhombs). Filled symbols correspond to the increasing branches of the applied voltage; empty symbols correspond to the decreasing branches. (Reprinted from Smerieri et al. [300], Copyright (2008), with permission from Elsevier)

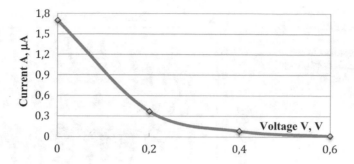

Fig. 6.4 Dependence of the max total current of the organic memristive device on the potential applied to the reference electrode

transferring into the reduced insulating state. The dependence of max total current on the value of the potential applied to the reference electrode is shown in Fig. 6.4.

When +0.6 V was applied to the reference electrode, the max conductivity of the device was decreased by more than two orders of magnitude with respect to the situation when the reference electrode was grounded. Moreover, when the complete cycle was finished, the resistance of the device reached the value of about 9 MΩ. It has resulted in the fact that it was necessary to perform additional doping of the device in HCl vapor for bringing it back to the conducting state.

Application of the negative potential to the reference electrode has not revealed such pronounced effects. In particular, the value of the max current was reduced by 20% when −0.2 V was applied, and for 50% when −0.6 V was applied.

Analysis of cyclic voltage-current characteristics for the ionic current revealed the shift of the peak position for the switching from the conducting state to the insulating one and vice versa. This shift is due to the fact that the actual potential of polyaniline in the active zone is shifted with respect to the potential applied to the reference electrode.

6.1.2 Pulse Mode

As was shown in the previous sub-section, even simple circuits based on a single memristive device, allow adaptations and training of such systems. The next step to mimic the living beings learning to mimic is connected to the study of the possibility of their training in a pulse mode.

Making again the parallel with a simplified model of Pavlov dog learning (disregarding timing of pulses that will be considered in the section, dedicated to STDP learning algorithm), we can attribute acoustic stimulus to the input voltage V_1 with respect to the basic level V_0, corresponding to the situation when the conductivity state of the device does not change. We will use two types of input signals

Fig. 6.5 Training of organic memristive device in a pulse mode [242]. (Reprinted from Smerieri et al. [242], with the permission of AIP Publishing)

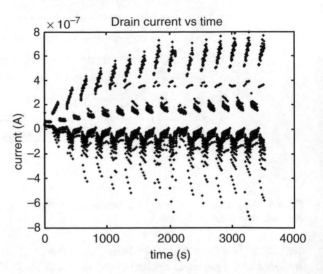

during the system training: positive training (presence of food), which must result in the reinforcement of the association, will be characterized by V_2; negative training (for example, when the acoustic signal is accompanied not by the presence of food but beating the dog) resulting in the inhibition of the association will be characterized by V_3. In the case of the positive training, pulses with an amplitude higher than V_1 will be applied, while in the case of negative training, pulses with negative potential will be applied. Similar to the DC mode, the value of the total current of the device will be considered as the output signal [242].

During the performed experiments, training and testing phases were alternated. During the training phase, we have applied a sequence of 5 pulses with the amplitude of $V_0 + V_2$ or $V_0 - V_3$, while during the testing phase, a sequence of 5 pulses with the amplitude of $V_0 + V_1$ was applied. The time between pulses was equal to the pulse duration. The time interval between different phases was usually 15 seconds. Values of V_0, V_1, V_2, and V_3 were chosen for each particular device after its careful characterization. Usually, these values were $V_0 = 0.3$ V, $V_1 = 0.1$ V, $V_2 = 0.3$ V, and $V_3 = 0.6$ V.

Experimental results when the organic memristive device was exposed to 15 cycles of training and testing are shown in Fig. 6.5.

The upper points in Fig. 6.5 correspond to the application of training pulses, points in the central part correspond to the testing pulses, and points in the bottom part correspond to the basic potential level. At the beginning of the experiment, application of the input, corresponding to the acoustic stimulus results in the value of the output signal of about 151 nA, which increases to 270 nA after the application of 10 training cycles.

Results of the experiment, when initially 30 testing cycles, followed by 15 cycles of training/testing were applied, are shown in Fig. 6.6.

Fig. 6.6 Temporal dependence of the output current when 30 testing cycles, followed by 15 pairs of training/testing cycles were applied to the organic memristive device. (Reprinted from Smerieri et al. [242], with the permission of AIP Publishing)

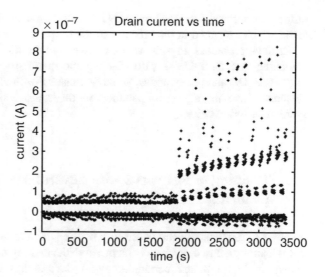

Fig. 6.7 Temporal dependence of the output current during the application of 15 reinforcing training/testing cycles (left part), 15 inhibiting training/testing cycles (middle part), and 15 reinforcing training/testing cycles (right part). (Reprinted from Smerieri et al. [242], with the permission of AIP Publishing)

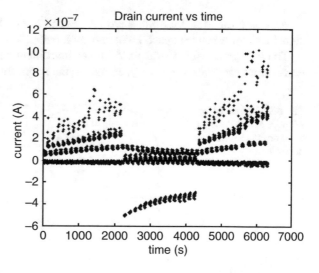

The left part of Fig. 6.6 represents two groups of experimental points corresponding to testing pulses and basic potential levels. The right part of this figure contains also the third group of experimental points corresponding to training pulses. As it is clear from the figure, application of testing cycles only does not vary practically the value of the output signal, while during the training phase, this value is continuously increased from 45 nA to 97 nA.

Experimental results, when reinforcing and inhibiting training cycles were applied, are shown in Fig. 6.7.

The experiment started with the application of 15 reinforcing/testing cycles. It has resulted in the fact that the value of the output current was increased from 45 nA up to 107 nA. After it, 15 inhibiting/testing cycles were applied, and the value of the output current was decreased up to 59 nA. Successive application of 15 reinforcing/testing cycles resulted in the increase of the output current up to 150 nA.

Results presented in this section have shown that the training (resistance switching in a desirable way) of the organic memristive device is very effective both in the DC and pulse modes, which indicates the possibility of the realization of systems where the signal propagation mode is more similar to that in nervous systems of living beings.

6.2 Training of Networks with Several Memristive Elements

In order to demonstrate the possibility of the adaptive network realization (supervised learning), based on organic memristive devices (synapse analogs), the network shown in Fig. 6.8 and containing eight devices has been realized and tested [247, 306].

As it is clear from Fig. 6.8, there are several possibilities for the formation of signal pathways between each input and each output terminal in this circuit.

The voltage of +0.6 V was used to bias both input terminals. Current values at each output electrode were analyzed during the application of the abovementioned

Fig. 6.8 Scheme of the adaptive network, based on eight organic memristive device

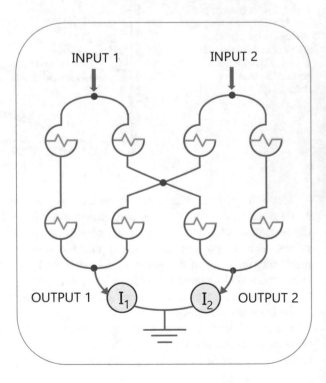

Table 6.1 Results of training of the circuit shown in Fig. 6.8

	Current at output 1 (nA)	Current at output 2 (nA)
Before training	120	32
After training	65	124

stimulus of +0.6 V to the first input electrode. Results of the measurements are reported in Table 6.1.

Initial values of output currents measured in points Output 1 and 2 are significantly different, which is due to the dispersion of properties and initial conditions of individual memristive devices. In addition, analysis of the distribution of elements (memristive devices) in the network indicates that the chain connecting Input 1 with the Output 1 must be more conductive than the chain connecting Input 1 with Output 2 (if we assume equal initial resistances of all memristive devices in the whole circuit).

As it is clear from Table 6.1, initially, the application of the input signal results in a higher value of the signal at the first output. Therefore, the training was done for the inversion of the situation: the application of the signal to the first input must result in a higher value of the signal at the second output terminal.

The training was done in the following way. The voltage of +1.2 V was applied between the first input and the second output for 5 minutes, while −0.5 V was applied between the first input and the first output. Testing of the system was performed after the end of the training phase. Similar to the case of testing before training, +0.6 V was applied to the first input and values of currents at Outputs 1 and 2 were measured. The results are presented in Table 6.1. As it is clear from Table 6.1, the training was successful and the conductivity of pathways was inverted. It is to underline that the variation of the conductivity of signal pathways, in this case, was done without any modification of the circuit but only due to the adequate training procedure.

The training of the system is a reversible process: applying adequately modified algorithms it was possible to bring the system to initial conditions or, if necessary, to suppress more the signal pathway between the first input and the second output.

It is to note that the use of organic materials allows additional possibilities to improve the functional properties of the system. In particular, the circuit shown in Fig. 6.8 has been realized in a flexible version. Kapton film was used as a support in this case. The image of the realized structure is shown in Fig. 6.9.

Cyclic voltage-current and kinetics characteristics of the system shown in Fig. 6.9, measured between any input and any output terminals, were absolutely the same as for the system in a rigid version.

Fig. 6.9 Adaptive network with eight memristive devices realized in a flexible version [306]. (Reprinted from Erokhin et al. [306], Copyright (2010), with permission from Elsevier)

6.3   Training Algorithms

Even if conventional training algorithms for traditional neural networks are rather well established, in the case of neuromorphic systems, they cannot be directly applied. In fact, even for the rather simple case of a double-layer perceptron, one needs to have the possibility of monitoring the weight functions of all connections and kinetics of their variation, as well as to be able to vary these weight functions performing necessary action to any memristive device (responsible for the weight function variation) in the network. It is rather easy to do with neuron networks at the software level: all elements are simulated and there is a possibility to access all of them (readout of the current resistance value and application of stimuli, capable to increase or decrease these values) without disturbing other elements. In nervous (and neuromorphic) systems, the situation is different. It is not possible to have an access to elements in intermediate layers and one can apply necessary voltages (or currents) only to input and output electrodes.

In this section, we will consider the training of the model circuit composed of 27 memristive elements. The training of the system, in this case, was the induction of a preferential signal pathway between a chosen pair of input-output electrodes and inhibition of the conductivity between all other possible pairs of electrodes.

The scheme of the model that will be discussed in this section is shown in Fig. 6.10.

The top panel of Fig. 6.10 demonstrates an equivalent circuit of the simplified model of a single organic memristive device considered in Chap. 3. Introduction of

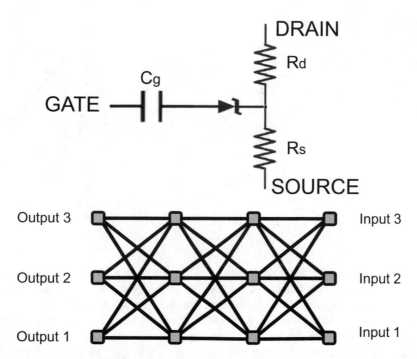

Fig. 6.10 Simplified equivalent circuit of a single organic memristive device (top panel) and scheme of the memristive network (bottom panel). (Top Figure reprinted from Erokhin et al. [306], Copyright (2010), with permission from Elsevier.)

Zener diode to the circuit is necessary for taking into account the variation of the device operation mode when reduction or oxidation potentials are reached.

The network shown in the bottom panel of Fig. 6.10 contains a layer of input nodes, two intermediate layers of nodes, and a layer of output nodes. Each layer contains three nodes. Each node of the layer is connected with all nodes in the previous and successive layers. Nodes are contact points, while connections between them are organic memristive devices. Thus, the network contains 27 memristive devices and 6 terminal nodes (three inputs and three outputs). Memristive devices are connected in such a way that their source electrodes are directed to output nodes, while their drain electrodes are directed to input nodes. It was supposed that each terminal node can be connected to a current generator, voltage generator, or zero potential. The configuration shown in Fig. 6.10 is an intermediate one between a single device and a complicated stochastic network, which will be considered in Chap. 7.

Modified nodal analysis method [307] was used for the calculation of parameters and evolution of the network.

For the formation of the conducting pathway between the chosen pair of input-output nodes, it is necessary to switch conductivity states of individual memristive devices of the chain by the application of appropriate voltage values to their

Fig. 6.11 Examples
illustrating situations when
it is impossible to make
training of the 2D network

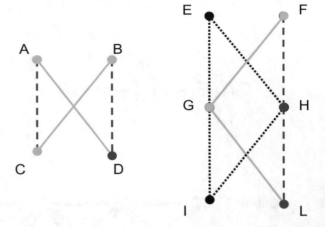

terminals. In the case when the memristive device is in a partially or completely insulating state, the application of the voltage higher than +0.4 V will result in the increase of its conductivity, while the application of the voltage lower than +0.1 V will result in the increase of its resistance. Thus, the variation of a single device conductivity is a very simple task requiring only the application of the adequate voltage for a sufficient time interval, while the realization of preferential signal pathways between the defined input-output pair of nodes is a much more complicated task, as it requires the increase of the conductivity of several memristive devices in the network with the decrease of the conductivity of all other devices.

If it would be possible to access all individual devices, the task would be rather trivial, as it is in the case of artificial neuron networks realized at the software level. However, in neuromorphic networks similar to the nervous systems of living beings, we can allow ourselves the application of training stimuli only to terminal nodes of the system. Thus, application of stimuli to the terminal nodes will be distributed among all memristive devices of the network.

In the case of 2D organization of the network, there are several restrictions on the realizability of the final distribution of conductivity of elements between input and output terminals during the application of stimuli to the whole system. Examples of such restrictions are shown in Fig. 6.11.

For example, in the case of the situation shown in the left part of Fig. 6.11, it is impossible to train the system in such a way that the conductivity between points A-D and B-C is high, while it is low between points A-C and B-D simultaneously (we are considering now 2D organized networks only; in the case of 3D organization this situation is possible, as it will be discussed in Chap. 7).

The right part of Fig. 6.11 illustrates the situation when the signal pathway involves more than one layer of memristive devices. In this case, for example, it is impossible to realize the situation when the chain F-G-L is in the conducting state, while the chain F-H-L is in the insulating one.

Taking into account the restrictions mentioned above, three training algorithms have been developed. For each strategy, a pair of input and output nodes was chosen, and the conductivity state between these nodes was checked at each moment of training. The difference between algorithms was connected to the choice of connections that must decrease their conductivity. In the case of the first algorithm, it was performed the decrease of the conductivity of connections, going from nodes within the signal transfer chain to nodes outside this chain. In the case of the second algorithm, it was performed the decrease of the conductivity of connections going from nodes outside the signal transfer chain to nodes within this chain. In the third case, the alteration of the two mentioned algorithms was applied.

During the practical application of all mentioned algorithms, we supposed that the distribution of the potentials among all resistive elements is the sum of all potentials applied to terminal nodes.

For each training algorithm, we suppose that some connections (memristive devices) must increase their conductivity, while the other ones must decrease it. Therefore, we put some limitations on nodes, providing these connections. For example, if the connection (memristive device) must come to a more conducting state, the potential difference on its terminal nodes must be more than V_{ox}. Thus, the application of the chosen algorithm implies the search of the potential distribution, satisfying all necessary conditions for this strategy.

On the other hand, any distribution of potentials in the network, resulted from the application of stimuli to six terminal nodes of the network, is a linear combination of the potential distribution on each node of the circuit. Moreover, as the variation of the conductivity state of each memristive device is determined by the voltage on its terminal electrodes, we can consider one arbitrary chosen node as a grounded one, and it can be considered also as a reference electrode.

It is possible to consider two stages during the application of the chosen algorithm to each particular network: determination of the preferential signal pathway and the application of the training procedure itself. First, we must identify nodes to be connected (one input and one output). After it, all possible pathways allowing these connections are considered, supposing that the pathway must contain only three memristive devices (one per layer). The distribution of the potentials between elements necessary for performing the training according to the chosen strategy is calculated for each possible pathway. After it, Output 1 was maintained at the ground potential level, and the distribution of the potentials between network nodes was calculated during successive variation of the applied current stimuli to all other terminal nodes. As a result, an array of five potential distribution profiles in the network was determined: $V_1 \ldots V_5$. After it, their linear combination, satisfying the signal transfer circuit parameters according to the chosen algorithm, was calculated. In the case when no one combination corresponded to the chosen algorithm, the other possible signal transfer pathway was considered. The training procedure was started when the combination $aV_1 + bV_2 + cV_3 + dV_4 + eV_5$, corresponding to the imposed parameters, was found.

Values of the applied training currents were equal to the product of unit currents on coefficients $a \ldots e$. Current with chosen values was applied to the system during the time interval dt. It was supposed that the values of the potentials on the electrodes of each memristive device were constant during this time interval. As the distribution of resistances was varied after each calculation step, calculation of new potential distribution $V_1,...,V_5$ and coefficients $a,...,e$ was done for satisfying the chosen conditions. Values of the currents for the next training step were taken according to the values of these new coefficients. The calculation process was finished if one of three conditions was reached: (1) There is no combination, improving the state of the system; (2) required values of currents are higher than predefine values (about ± 10 mA, this value will result in the damage of the organic memristive device); (3) iteration number is higher than the predetermined value.

The efficiency of the network training was estimated according to four criteria. The first two, called "reinforcement" and "threshold," were connected to the influence of the values of the applied currents on the formation of input-output signal pathways.

Let us consider how the current I_{AB} between Input A and Output B will be varied when Output B is at ground potential and the potential of 1.0 V is applied to Input A. For example, if our task was the induction of the conducting pathway between Input 3 and Output 2, the values of "reinforcement" (G) and "threshold" (O) will be determined by Eq. 6.1.

$$G \equiv \min\left(I_{32}/I_{31}, I_{32}/I_{33}\right),\ \ O \equiv \min\left(I_{32} - I_{31}, I_{32} - I_{33}\right) \tag{6.1}$$

The value of G is more than one, and the value of O is more than zero when the value of the current at the chosen output is more than at two other outputs. These values indicate how strong and selective is the connection between the chosen output with the chosen input.

Two other parameters, called "inversed reinforcement" (R) and "inversed threshold" (S), indicate the efficiency and selectivity of the signal pathway formation, analyzing the value at the chosen output electrode when stimuli are applied to the chosen and two other input nodes. These values are determined by Eq. 6.2.

$$R \equiv \min\left(I_{32}/I_{12}, I_{32}/I_{22}\right),\ \ S \equiv \min\left(I_{32} - I_{12},\ I_{32} - I_{22}\right) \tag{6.2}$$

Usually, for the arbitrary built system, values of G and R are slightly less than one, and values of O and S are slightly less than zero.

In the future, we will consider mainly the value of the "inversed reinforcement" R, because this value does not depend on the value of the signal applied to the input nodes. However, in some cases, it will be necessary to consider also values of "threshold" O and "inversed threshold" S, when absolute differences of the conductivity values will be more important than their ratios.

Each strategy was applied 1000 times to a randomly chosen system.

Step-by-step results of the application of each system are shown in Fig. 6.12.

As it is clear from Fig. 6.12, the final state of the system is very similar after the application of strategies with diverging and converging connections. The result is significantly different in the case of the application of the alternating strategy. In

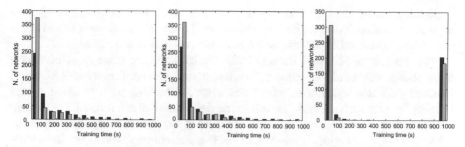

Fig. 6.12 Distribution of the required training time for three described strategies during 500 successive training steps. Left panel – the strategy of diverging connections, middle panel – strategy of converging connections, right panel – alteration of diverging and converging connections. Dark columns correspond to the case, when all range of resistances was possible (50 kΩ–1 MΩ), green columns correspond to reduced resistance range (500 kΩ–1 MΩ)

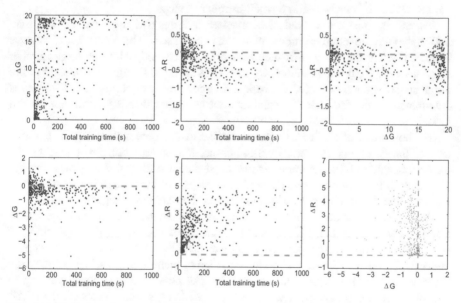

Fig. 6.13 Variation of parameters G and R during the training process. Each point represents the variation of the individual network, considering that the total number of networks was equal to 500. Top row: training with the strategy of diverging connections; bottom row: strategy of converging connections

about a half of cases, systems required training till the maximum allowed number of steps, while in the rest of cases it was stopped after 150 steps (even if for most of them the training was finished after 20 steps). Instead, for strategies with diverging and converging connections, we can observe a fast decay of required time. Only a small part of the networks was subjected to more than 100 steps of training. In the case of using the reduced resistance range, fewer steps were required for training.

Results of training are shown in Fig. 6.13.

In this case, we have used "reinforcement" and "inversed reinforcement" parameters for the estimation of the training efficiency. Similar results were obtained in the
case when "threshold" and "inversed threshold" parameters were used.

The top row of Fig. 6.13 demonstrates results when the strategy of diverging
connections was used, showing the difference of the reinforcement (ΔG) and
inversed reinforcement (ΔR), before and after the training as a function of the
training time for each network, as well as the dependence of ΔR on ΔG. The bottom
row demonstrates similar dependences, obtained in the case of the application of the
strategy with converging connections. Each point corresponds to an individual
network chosen from 500 randomly built networks after training. In this case, the
whole range of possible resistance values was allowed.

It is to note that there is some symmetry in the distribution of results during the
application of diverging and converging strategies. In the case of the first strategy,
we can see the increase of the reinforcing with time, even if the maximum difference
(about 20) can be reached after a small number of steps.

Instead, in most cases, (and always when training was done during less than
300 steps) the inverted reinforcement was decreased. The value of G always
increases, while the value of R increases only for several points. A large number
of points near (0,0) is determined by the fact that the training was finished at the
initial stage for the most tested networks. A large number of points near $\Delta G = 20$ is
determined by the fact that this value cannot be higher than 20 within the chosen
model.

The bottom panel of Fig. 6.13 shows similar results obtained during the application of the diverging strategy. In most cases, R increases but never reaches the
maximum value of 20. The presence of cluster near (0,0) point indicates that the
training was finished after a small number of steps. An increase in the training time
results in the increase of ΔR value and the fact that ΔG has a negative value.

Results obtained during the application of the alternating strategy are rather
different and unexpected, with respect to results obtained during the application of
previously mentioned strategies. As it is clear from Fig. 6.14a, showing the dependence of ΔR on ΔG, there are two clusters. One of them is at the initial point, while
the other one is located near the point (4,2). The most of points corresponding to
calculation results are concentrated in the first quadrant, which reveals the improved
reinforcement and inverted reinforcement when this strategy is applied. Moreover,
even if we will consider networks when the training was carried out till the
predefined limit, the two mentioned clusters will be still observed (Fig. 6.14b).
According to Fig. 6.14, the dispersion of training results is more pronounced when
we consider the whole range of possible variation of resistances, with respect to the
restricted range (500 kΩ–1 MΩ).

Figure 6.15 shows the evolution of the state of 1000 networks after the application of the alternating training procedure supposing the whole possible resistance
range and limit of 5000 training steps. For better comparison of the final result,
points on the dependence correspond to real values of G and R, and not to the
differences before and after training (Fig. 6.15 right part).

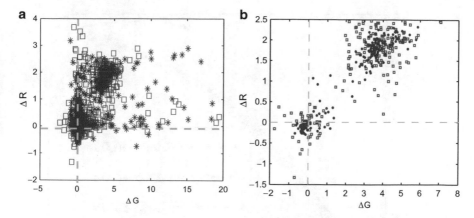

Fig. 6.14 Dependence of ΔR on ΔG when alternating training strategy was applied for 1000 networks. (**a**) All networks; (**b**) Networks, for whom the training was not finished before the pre-imposed time limit. Squares correspond to training, allowing all possible resistances; stars correspond to networks, where the resistance range was limited to 500 kΩ–1 MΩ

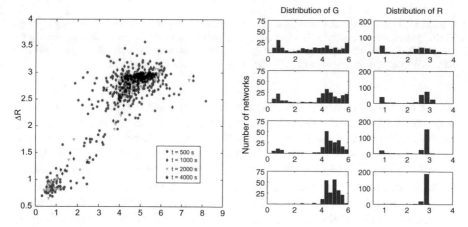

Fig. 6.15 (Left) Dependence of ΔR on ΔG for 1000 randomly built networks after the application of alternating training procedure. (Right) Dependences of R and G parameters for these networks in different moments of training. Histograms show the distribution of G and R values for 500, 1000, 2000, and 4000 training steps

As was in the previous case, the training of networks was or rather fast (less than 200 steps), or it was continued till reaching the imposed time limit. Figure 6.15 shows the result of the second case. For most networks, the reaching of the maximum area that is located near the point (5,3) in Fig. 6.15a took place within 500 steps of training, if initial values of G and R were close to (1,1) point. After 4000 training steps, practically all points form a cluster near the point (5,3).

Figure 6.16 shows the dependence of ΔR on ΔG according to the evolution of a single network during training. During the first 3000 steps, the system is in a first

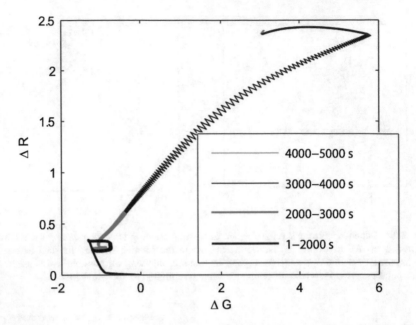

Fig. 6.16 Dependence of ΔR on ΔG of a single network during training with alternating strategy. Training starts at the point (0,0). The first 2000 steps are represented by a bold solid line; steps in the range of 2001–3000 are shown as a bold grey line, steps in the range of 3001–4000 are shown as a thin solid line, steps in the range of 4001–5000 are shown as a thin gray line

cluster where the value of ΔR is less than 0.5 and ΔG is negative. Successive 1000 steps shift the system state towards the next domain. The next 1000 steps do not vary practically the state of the system. The effect of the application of the alternating training procedure is very pronounced in the range of 3000–4000 steps. The training curve is a zigzag line, where each next step changes the direction to the opposite one. Thus, the displacement due to two steps is significantly smaller than for one step.

Let us recall that this section is dedicated to the development of training strategies of networks of memristive elements with stochastically distributed properties. Even if such networks represent simplified cases, they allow to consider several important particularities that will be present in more complicated stochastic networks, such as inter-layers of memristive devices where direct access is forbidden and the impossibility of simultaneous adaptation of connections. Three training strategies were developed, and parameters of their efficiency were determined. It was also suggested a method of the determination of correct stimuli values to terminal nodes of the system for the execution of effective training according to the identified strategy. The method is based on the application of the superposition principle taking into account threshold values of the resistance switching of memristive elements.

It is to note that the developed approach is applicable not only to hypothetical circuits but can be also applied to real networks. Ideally, it can be done if we can reconstruct the map of the resistance distribution within elements of the network, which can be done as was discussed in Chap. 3, using the optical imaging of the whole network during its functioning.

Both considered strategies, namely, based on diverging and converging bonds, make symmetric action on the network. Each of them improves the value of one parameter, while the other one becomes worth. Both strategies are more efficient in the reduced range of possible resistances. It seems that the application of these strategies will be very efficient in cases when the value of one efficiency parameter is already rather high, and it is necessary to improve just the value of the other parameter.

In the case of the third strategy, that is the combination of the first two, completely other behavior has been observed. First, as it has been shown, this strategy has divided all randomly generated networks into two classes. Networks, corresponding to the first class, have finished their training within less than 200 steps, while those corresponding to the second class, continued their training at least 20 times longer.

Two efficiency estimating parameters were improved for both classes. However, for the second class of networks, the final values of R and G parameters were rather close to each other disregarding the initial state of the network. During the initial stage of training, the variation of the system state occurs near the (1,1) point (Fig. 6.16). During the training, the state of the system is shifting to the point (4.3). If the training is rather long, practically no points remain in the first cluster. Thus, for this class of networks, their initial state is practically fixed. However, at a certain moment, we observe a fast transition of the network to its final state, where a significant improvement of both efficiency parameters can be observed with respect to their initial values.

6.4   Electronic Analog of the Part of the Nervous System of Pond Snail (Lymnaea Stagnalis)

Last years, a lot of attention is dedicated to researches directed to the utilization of living beings' organization principles during the design of new materials and devices. In the case of information processing, we can see a significant increase in research activity connected to the expected breakthrough after the utilization of memristive devices for mimicking synapse and neuron basic functions. It will allow to make electronic circuits and networks capable of learning at the hardware level. Currently, it appeared also a new term – "neuromorphic systems."

Synapse properties can be realized also using traditional electronic compounds [308–320]. However, in the case of using traditional electronic compounds, a rather large number of devices is required. Instead, at least for synapse mimicking, it can be done at the level of single elements when memristive devices are used. Thus, the use of memristive devices, in general, and organic memristive devices, in particular, will allow to simplify significantly the architecture of neuromorphic circuits, to increase the integration level, to decrease power consumption, and, as a result, to diminish the probability of errors. In addition, organic materials allow self-organization, which can lead to the formation of neuromorphic networks with 3D architecture, which is absolutely impossible when working with inorganic materials.

In this section, we will consider the first direct demonstration of the fact that organic memristive devices can work as a synapse analog in electronic systems. For this reason, we will consider a circuit mimicking the architecture of the part of the nervous system of a pond snail, *Lymnaea stagnalis,* responsible for learning of the animal during feeding. The position of organic memristive devices in the circuit corresponds to the position of synapses in the nervous system of the animal.

The choice of the pond snail as a biological benchmark was due to the fact that its learning was studied in detail, which resulted in the development of the mathematical model of the electrophysiological basis of the plasticity and reconstruct the architecture of the part of the nervous system [158–163].

6.4.1 Biological Benchmark

In the case of the snail, learning means the association of the initially neutral conditioned stimulus (CS) (touching of the snail's mouth) with the presence of food, which leads to a rhythmic contraction of the animal muscles necessary for swallowing and digestion of the food. Before learning, such contraction can be observed only in the case of the food presence (unconditioned stimulus (US)) but never in the presence of CS only. Learning of the snail occurred when both stimuli were applied simultaneously, which resulted in the starting of muscle contraction after successive application of CS only. In other words, the animal has learned to associate lip touching with the presence of food [321].

Experimental neurophysiological results allow to distinguish two types of synaptic plasticity: homo-synaptic or Hebbian plasticity [7] and hetero-synaptic or modulatory plasticity [322, 323]. It has been suggested that homo-synaptic plasticity is responsible for learning and short-term memory, while the hetero-synaptic one is responsible for maintaining the learning results and for the long-term memory [324].

The considered part of the nervous system of *Lymnaea stagnalis* can be divided into functional groups of neurons [321]: sensor neurons, intermediate modulatory neurons, intermediate neurons of Central Pattern Generator (CPG), and actuating neurons. Cerebral giant cells (CGCs) are a very important part of the network connecting pairs of sensors, intermediate, and actuating neurons. It has been shown that CGCs play a very important role in the formation of long-term memory during associative learning. In particular, after a single learning event, a depolarized potential appears at the CGCs membrane which explains the increased response of the digesting system of the animal as the result of CS application.

We will consider this hetero-synaptic interaction in this section. In particular, in biological systems, the strength of synaptic connections between two neurons can be varied by the application of a modulatory signal to the third neuron. Thus, also the electronic circuits must mimic similar behavior. The scheme of the part of the

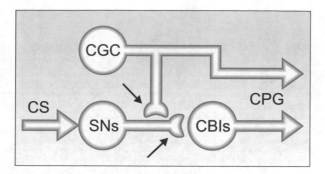

Fig. 6.17 Diagram showing the interaction of cerebral giant cells (CGCs) with sensory (SNs) and command intermediate (Cerebro-buccal interneurons (CBIs)), which mediate the increased response of the system to the conditioned stimulus (CS) after conditioning. Arrows show the position of synapses [156]. (Reprinted by permission from Springer Nature, Erokhin et al. [156], Copyright (2011))

nervous system of the pond snail responsible for learning during the application of CS is shown in Fig. 6.17.

In the case of the architecture shown in Fig. 6.17, the presence of only one input stimulus corresponding to CS results in the output signal value, which is lower than the threshold necessary for starting the digestion process. After learning, the properties of the system are changed in such a way that the application of CS is enough for having an output signal with an amplitude higher than the threshold value.

6.4.2 Experimentally Realized Circuit, Mimicking the Architecture and Properties of the Pond Snail Nervous System Part

Considering the reasons mentioned above, two types of electronic circuits were realized corresponding to homo- and hetero-synaptic types of learning. Schemes of the realized circuits are shown in Fig. 6.18a, b, respectively, while experimental results are shown in Fig. 6.18c, d [156].

Similar to the real animal, both schemes shown in Fig. 6.18 have two inputs corresponding to the application of neutral stimulus (touching – in 1) and presence of food (in 2). A sinusoidal signal with the frequency of 1 Hz, amplitude of 0.1 V, and offset of +0.3 V was used as the neutral stimulus. The necessity of the offset is due to the fact that the reduction potential of polyaniline is about +0.1 V. Thus, the chosen values of the amplitude and the offset guaranteed the constancy of the conductivity state of memristive devices in the circuits. For having more similarities to living beings, AC signal was used in this case. However, it is not so critical for the design of electronic circuits allowing learning processes. Let us consider this item in more detail.

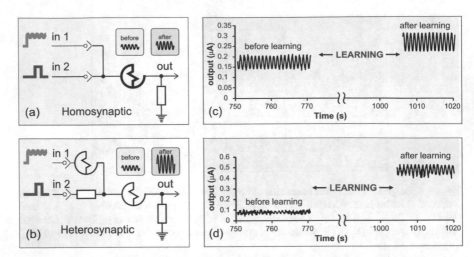

Fig. 6.18 Schemes of electronic circuits, mimicking homo- (**a**) and hetero-synaptic (**b**) learning of pond snail *Lymnaea stagnalis*. (**c**) Experimental results on learning of circuit (**a**); (**d**) Experimental results on learning of circuit (**b**). (Reprinted by permission from Springer Nature, Erokhin et al. [156], Copyright (2011))

Living beings are complicated systems performing a complex set of functions. Therefore, the presence of similar mechanisms and properties combining individual components in a unique system is required. Namely, the energy used for the functioning of all elements must have the same nature; sensory elements, the information processing system (nervous system), and executive organs must be compatible with each other. Nature has chosen organic materials capable to vary their properties according to electrochemical redox reactions as building materials for living beings. However, redox reactions demand the directed flow of ions, including processes in the nervous system. As the signal propagation in the nervous system implies anisotropy, the use of pulse mode in nature is absolutely a necessary condition. Otherwise, it could result in the appearance of gradients of ions that will block further information processing. In the case of artificial electronic systems, based, in particular, on organic memristive devices, this is, in principle, not necessary. At least in the current moment, it is impossible to realize a system that will include elements allowing to collect and process the information, as well as executive elements, having, moreover, the same nature and comparable signal values. Thus, considering the current technological level, we can expect the realization of hybrid systems including different sensory elements, information processing systems, allowing learning and decision-making, and executive elements. Each particular realization of each element will be based on the current technological possibilities, and their combination into a unique system will require the use of adaptive intermediate devices.

In the present work, we consider only bio-mimicking information processing systems. Even if the working principle of organic memristive devices, similar to that in living beings, is based on redox reactions, the area where they occur is restricted by the active zone and the direction of ions motion is perpendicular to that of the signal propagation. Therefore, the information processing will not result in the appearance of the ion concentration gradients on the length of the signal transfer pathways. Therefore, blocking of the information processing will not take place. In summary, it is possible to claim that neuromorphic networks based on organic memristive devices can work also in DC mode. Nevertheless, pulse mode seems to be an unavoidable condition for the realization of unsupervised learning according to spike-timing-dependent plasticity (STDP) that is considered in appropriate sections.

Let us come back to the results shown in Fig. 6.18. As was mentioned above, a signal with the frequency of 1 Hz was applied to Input 1. The value of the frequency is not important, and it can be increased or decreased, which will not change the final characteristics of the system.

In the case of homo-synaptic architecture, the experiment was done in the following way. The neutral signal was applied to the first input for 10 minutes with continuous monitoring of the output signal value. The training of the circuit was done during 15 successive minutes when the signal corresponding to the presence of food was applied to Input 2, while the signal on Input 1 was also maintained. The value of the signal applied to Input 2 was 0.3 V. When the training phase was finished, the signal was applied again only to Input 1. Values of the output current were compared for the measurements before and after the training phase.

In the case of hetero-synaptic architecture, schematically shown in Fig. 6.18b, a neutral sinusoidal signal with the frequency 1 Hz and amplitude of 0.1 V was also applied to Input 1, but its constant offset was equal to +0.6 V. The increased value of the offset, in this case, is due to the fact that this voltage is distributed between two memristive devices. As was in the previous case, the training phase, involving the application of the signal of +0.6 V to the second input for 20 minutes, was started 10 minutes after the beginning of the experiment. After finishing the training phase, the initially neutral signal was applied again only to the first input, and the obtained value was compared to that obtained before the training. Results of learning of homo- and hetero-synaptic circuits are shown in Fig. 6.18c, d, respectively. It is to note the difference in the scale of output currents and rather high level of noise before the training.

First, the simplest circuit shown in Fig. 6.18a was tested. The choice of values of input signals was done in such a way that their successive application must not result in the variation of the resistance state of the memristive device. Therefore, the value of the output signal was just proportional to the applied signal according to the Ohm law. When both signals were applied simultaneously, a potential difference on the memristive device was enough for its switching to a more conducting state. Temporal dependences of the output signal of this circuit before and after training, when only a neutral stimulus was applied, are shown in Fig. 6.18c. As it is clear from the figure, the training was successful, which has resulted in about a 50% increase of the amplitude of the output current and a shift of its DC offset value.

The scheme of the network allowing hetero-synaptic learning is shown in Fig. 6.18b. Its architecture is very similar to the model of the part of the pond snail nervous system shown in Fig. 6.17. The position of organic memristive devices corresponds directly to the position of synapses in the model. The value of the signal, applied to Input 1 was such, that it can provide the resistance switching of only one memristive device. However, as this voltage is distributed between two devices, its amplitude is not enough for transferring no one of them into a more conducting state. During the training phase, the voltage was applied to the second input and its amplitude was enough for switching one of the memristive devices (positioned closer to the output) in a conducting state. When it took place, the DC compound of the signal on Input 1 was mainly applied to the first memristive device, transferring it to the conducting state. Results of the training of the circuit with heterosynaptic architecture are shown in Fig. 6.18d. As it is clear from the figure, training of the system resulted in a fivefold increase of both amplitude and offset of the output signal.

The described behavior corresponds well to the modulatory role of heterosynaptic junctions in biological neural networks and confirms the hypothesis of their role in the formation of long-term memory [324]. Moreover, this modulatory input can trigger a cascade of intramolecular events, which lead to relatively long-term modifications of synaptic function. Specifically, while activation of the homosynaptic pathway (SNs → CBIs, Fig. 6.17) by the CS does not lead to a significant response, simultaneous activation of the modulatory pathway (CGCs → SNs; Fig. 6.17) persistently facilitates the synapse (SNs → CBIs, Fig. 6.17) and allows for an increased response of the circuit to the CS only. Similarly, in the realized synthetic circuit, activation of the homosynaptic pathway by the CS analog only leads to a relatively weak response, while the activation of both the homosynaptic and heterosynaptic pathways by the CS and US analogs, respectively, conditions the circuit so that a stronger response is triggered by the CS analog only.

Kinetics of the output signal variation during the learning phase (application of both input signals) is shown in Fig. 6.19 (time scale corresponds to the break interval in Fig. 6.18d).

The interesting feature of the dependence shown in Fig. 6.19 is a small decrease of the output signal value during the initial stage, followed by its significant increase in the following moments. This behavior is connected to the dispersion of properties of individual memristive devices in the circuit. The difference in the initial resistance values of different elements of the circuit results in the fact that the applied voltage is mainly distributed on the higher resistive element (this item will be considered in more detail in the successive section, dedicated to cross-talk of memristive elements). Therefore, switching to the more conducting state will occur first on the memristive device with higher resistance. This switching will redistribute the potential profile on the chain of elements which can result in the fact that the voltage on this element will be less than +0.1 V and it will increase the resistance due to the reduction of polyaniline in the channel. This effect is responsible for the approximate constancy of the circuit resistance till the moment when resistances of both elements will be comparable. After it, both of them will switch to a more conducting state.

The results discussed in this section are very important also for more complicated networks composed of a large number of stochastically distributed analogs of

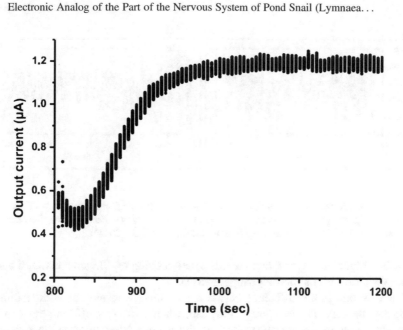

Fig. 6.19 Temporal variation of the output signal of the circuit, shown in Fig. 6.18b, during the training phase. (Reprinted by permission from Springer Nature, Erokhin et al. [156], Copyright (2011))

memristive devices, capable to form different signal pathways according to the learning algorithms, connected to the frequency, amplitude, and/or duration of the allied stimuli. During the learning procedure, the system will react in such a way that occasional variations of connections, resulted from short-term not repeatable combinations of external stimuli, will be compensated, and properties of the system will be significantly changed only in the case of the presence of long-term and/or repeatable tendencies. Moreover, as was discussed in Chap. 3, in particular conditions, the system can come to the auto-oscillation mode, which will maintain the system in dynamic equilibrium. This behavior is rather similar to processes in the nervous system and brain, which makes them ready to react to the new coming information and to provide its treatment taking into account the previous experience of the system connected to the formed short- and long-term connections between elements of the network. It is to underline that in this respect, such systems will be more similar to the nervous system than to the traditional computer memory. In fact, traditional memory is a passive system allowing recording, readout, and cancelation of the information. The presence of the cross-talk between memory units is considered a very big drawback for traditional computer memory. Instead, such behavior is typical for the nervous system and brain. Moreover, cognitive processes would not be possible without such cross-talk of nervous cells. Cognitive processes can be connected to the reinforcement/inhibition of different signal pathways according to the analysis of external stimuli arrays not necessary at the moment of their acquisition. Thinking does not require new stimuli; probably, this process implies the

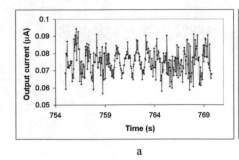 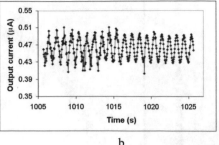

a   b

Fig. 6.20 Zoomed version of results, shown in Fig. 6.18d: (**a**) temporal dependence of the output signal before learning; (**b**) temporal dependence of the output signal after learning. (Reprinted by permission from Springer Nature, Erokhin et al. [156], Copyright (2011))

adequate choice of network regions for superimposing on them new stimuli and to form new associations (new signal pathways).

There is one more difference between the considered system with traditional computer memory. For computers, the case, when switching off the power supply does not result in the transformation of the accumulated information (so-called nonvolatile memory), is considered as an ideal characteristic [325–333]. In fact, as was mentioned above, a passive role of the information in modern computers implies a constancy of accumulated data without the external command. Thus, the performance of the modern computer (at least at the hardware level) is determined by the programming and current configuration of input signals. The system based on organic memristive devices is closer to the nervous system. Each its current state is based on the whole accumulated experience of its functioning and the internal activity based on cross talk of elements. Therefore, switching off of the power supply will result in the loss of all accumulated experience and we will work with a different system with respect to that before switching off (in the case of short-term switching off the power supply, "strong" signal pathways will be maintained—the system will save main properties—while short-term associations will be "forgotten"). It is to note that similar considerations were done also for other volatile memristive systems [334–337].

The above considerations allow to understand better the dependence shown in Fig. 6.18d. In the case of homo-synaptic learning, we can observe only a small increase of the amplitude and offset of the output current value, while in the case of the hetero-synaptic junction, we can see also a decreased level of noise, as is clear from Fig. 6.20.

As it is clear from Fig. 6.20, sinusoidal output signal before learning has a rather high level of noise, even if its sinusoidal character is still obvious. This behavior is significantly different with respect to the case of the homo-synaptic circuit shown in Fig. 6.18a, which is due to the fact that in the case of the hetero-synaptic junction, we deal with a cascade composed of two memristive devices. Complex nonlinear character of functioning of each memristive device results in the fact that the noise level increases with the number of unbalanced devices in the signal transfer chain.

However, after learning, this noise level was significantly decreased. In addition, this noise level continues to decrease during the circuit operation. This behavior demonstrates that the network of organic memristive devices is a system capable to adapt properties of individual elements of signal pathways during the operation in such a way that the noise level of the system decreases.

Further considerations on the noise effect to organic memristive devices will be considered in the appropriate section.

6.5   Cross Talk of Memristive Devices During Signal Pathways Formation Process

In this section, we will consider circuits containing more than one organic memristive device in the signal transfer chain. In particular, we will consider circuits composed of two and three organic memristive devices [338, 339], as is shown in Fig. 6.21a, b, respectively.

In the case of the circuit composed of two memristive devices (Fig. 6.21a), measurements were started from the initial voltage of 0 V and increased with a step of 0.1 V till the maximum value of +2.0 V. After reaching the max value, the voltage was decreased with the same step till the max negative voltage of −2.0 V and then again increased to close the cycle at 0 V. A delay of 60 s was used after the application of each new voltage value before the readout of the current and application of the next voltage.

Temporal dependences of the total and ionic currents for the two elements circuit are shown in Fig. 6.22.

The shown characteristics have revealed some particularities, which were never observed in circuits with one organic memristive device only, described in detail in Chap. 2. In the positive branch of the applied voltage, there are two peaks both of the total and ionic currents. In the negative branch, two peaks are clearly visible only for the ionic current, which is due to the significant difference in the total resistance of the device in its conducting and insulating states.

In the case of the circuit composed from three memristive devices, measurements were done similarly with the only difference that max applied voltage was 3.0 V. The experimentally acquired dependences are shown in Fig. 6.23.

In the case of ionic current, the characteristic has even more complicated behavior with respect to that for the circuit composed of two devices, but in the case of the total current, it has a more smooth shape, having even a plateau. The cyclic voltage-current characteristic of this sample is shown in Fig. 6.24. Current values shown in Fig. 6.24 were measured at the end of each time interval after the application of successive voltage values, similar to the cases described in Chap. 2 for individual organic memristive devices.

The behavior shown in Fig. 6.24 can be explained in the following way. The applied voltage is distributed on all elements of the circuit. Switching of each

Fig. 6.21 Schemes of circuits containing two (**a**) and three (**b**) organic memristive devices in the signal transfer pathway. In both cases, the voltage was applied between S and D electrodes, and ionic current was measured between D and G1 electrodes. (Reprinted from Berzina et al. [338], with the permission of AIP Publishing)

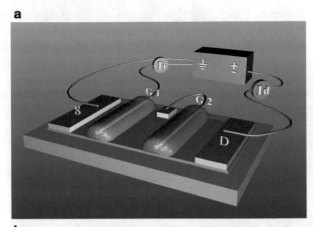

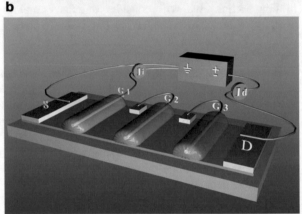

element into conducting (insulating) state occurs when its active zone (contact of polyaniline with polyethylene oxide) reaches oxidation (reduction) potential with respect to the reference electrode. It determines successive resistance switching of the circuit elements. The applied voltage is mainly concentrated on elements with higher resistance. Therefore, these elements (in the case of the positive applied voltage) will be the first ones that will reach the oxidation potential and will be transferred into a more conducting state. However, when one element in the chain becomes more conducting, this will result in the redistribution of the potential, which will bring other elements to the conditions, allowing their switching to a more conducting state, while the mentioned elements will be at a reduction potential and, therefore, will increase their resistance. Processes, in this case, have some similarity with those when the memristive device work in the auto-oscillation mode considered in Chap. 3.

In summary, it is possible to state that there are cross talk processes between organic memristive devices during the formation of stable signal pathways. It

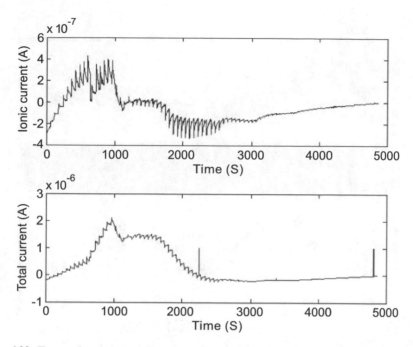

Fig. 6.22 Temporal variation of the ionic (upper panel) and total (lower panel) currents of the sample, composed of two serially connected organic memristive devices during cyclic application of the voltage scans between S and D electrodes. (Reprinted from Berzina et al. [338], with the permission of AIP Publishing)

demonstrates once again that the properties of such systems are more similar to the nervous system and not to traditional computer memory. The formation of stable signal pathways requires the balance of properties of constituting elements. Moreover, such a system will provide "self-repairing" properties: occasional variations of the conductivity of individual constituting elements will result in the redistribution of the potential along the whole chain, causing the variation of the conductivity of these elements till resistances of all elements (organic memristive devices) of the chain will be balanced.

6.6   Effect of Noise

In simple circuits and systems, the role of the noise is always negative. However, when the complexity of the system is increased, it can play also a positive role [340–342], bringing, for example, the system to a deeper stable energetic level, as it occurs in biological samples [343, 344].

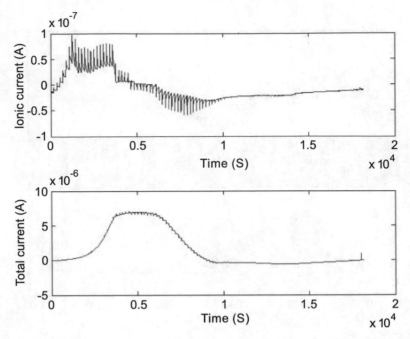

Fig. 6.23 Temporal variation of the ionic (upper panel) and total (lower panel) currents of the sample, composed from three serially connected organic memristive devices during cyclic application of the voltage scans between S and D electrodes. (Reprinted from Berzina et al. [338], with the permission of AIP Publishing)

Even if the noise effect is very important also in the field of circuits and systems, based on memristive devices, there are not so many publications dedicated to this item [345–351].

It is reported in [352] that the combination of the white noise of appropriate intensity even at very low frequencies and an external driving field could induce modifications in the hysteresis of memory elements. This study conducted using a reference device for modeling inorganic memristors, based on vacancy migration, anticipated the experimental demonstration reported in [346, 353] and a second simulation in which authors concluded that the charge probability density function may be modified in memristive circuits coupled to white noise sources.

It has been reported that the response of a memristor based on a $ZrO_2(Y)/Ta_2O_5$ stack to a white Gaussian noise signal demonstrating a stochastic switching in the random telegraph signal mode [349]. It has been proposed the use of noise as a tool for understanding the mechanisms of the resistive switching in memristors [354]. A further model was proposed in 2015, by Georgiou et al. [355], in which authors provided a simulation of different circuits based on an ideal memristor under the influence of thermal noise. Besides their results on the possible variation in memristive figures of merit on the basis of the circuit used, authors provided simulation using the classical external noise source in which the voltage is constant

Fig. 6.24 Cyclic voltage-current characteristics for ionic (upper panel) and electronic (lower panel) currents in the circuit, composed from three organic memristive devices (electronic current was calculated as the difference between total and ionic currents values). (Reprinted from Berzina et al. [338], with the permission of AIP Publishing)

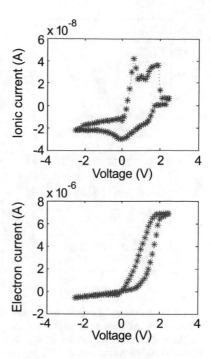

and where the amplitude is expressed as a function of the memristance of the device. From a more perspective point of view, the application of complex compound waveforms, in which a signal contains instantaneously a broad frequency range, is advantageous for impedance analysis (i.e. impedance sensing of biological systems) [356]. In fact, organic electrochemical transistors (OECTs) response has been tested delivering the noise signal from the gate. The noise was demonstrated to be particularly beneficial in monitoring the impedance of cultured epithelial monolayers in real time.

In this section, it will be reported the effects of white noise stimulation on organic memristive devices realized with a liquid electrolyte, as was described in Chap. 2, to ensure a fast resistive switching. These results suggest that the application of a voltage sweep with different noise amplitude levels, limiting the frequencies band to the brain occipital alpha rhythm (~10 Hz), alters significantly hysteresis that was quantified introducing a new performance parameter for organic memristive classes of devices [357].

As was already discussed in Chap. 2, the replacing of the solid polymeric electrolyte with a liquid one leads to higher stability (more than 250 consecutive cycles) of the performances and a faster switching time, as is shown in Fig. 6.25a (black solid line), by the formation of a wide hysteresis loop in the output current I_{SD}. During the voltage sweep, OMDs switch conductivity at around 0.4 V passing from a high to a low resistivity state, and in the opposite direction between 0.1 and 0.2 V.

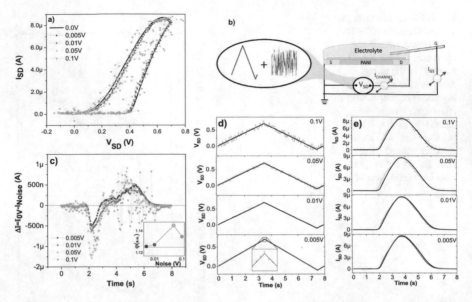

Fig. 6.25 Response of organic memristive device to noise sourcing: (**a**) I-V hysteresis behavior as a function of different noise amplitudes. (**b**) Scheme of the device and electrical connections. (**c**) ΔI vs Time. (**d**) Temporal evolution of V_{SD} for different noise amplitudes plotted with respect to the no noised stimulation (black line). (**e**) Temporal evolution of I_{SD} for different noise amplitudes plotted with respect to the no noised stimulation (black line) [357]. (Reprinted from Battistoni et al. [357], Copyright (2020), with permission from Elsevier)

When a constant amplitude noise contribution is superimposed to the voltage, applied between the source and drain electrodes (as depicted in Fig. 6.25b), organic memristive devices preserve the hysteretic behavior (reported in Fig. 6.25a) but reduce the width of the loops. This compression is not typical of consequent measurement cycles that produce a rigid shift of the entire curve to higher voltage values.

The entity of compression seems to follow the amplitude of the noise, and it is the result of the shift to lower voltage values of the first part of the sweep responsible for the transition to a high conductivity state and a translation to higher values in the back scan as shown in Fig. 6.25d, e. To better visualize this effect, we introduced the parameter $\Delta I = I_{0V} - I_{noise}$ as the performance parameter, and it is reported in Fig. 6.25c the temporal evolution of ΔI for different noise amplitudes. By definition, a dotted line at 0 A corresponds to the ΔI for the noiseless curve. The application of a noised source leads to the generation of two peaks in ΔI profiles, a negative one localized between 2 and 3 s (corresponding to 0.4 V and 0.6 V) and a broader and positive one corresponding to the back scan. The presence of the negative peak allows to distinguish efficiently noise and noiseless stimulation, and the qualitative description of the contraction of the hysteresis loop (ΔI) is particularly effective in the representation of the internal switching dynamics of the organic memristive devices. It is to note that peaks in Fig. 6.25c are centered in correspondence of the oxidation and reduction processes suggesting that the superimposition of the noised signal contributes to accelerating redox reactions in PANI.

These results are in good agreement with the general conclusion of references [346, 349, 353]: noise alters the switching process and the width of the hysteresis loop. In our case, however, we did not observe an enhancing of the hysteresis loop but rather an increase in the probability of redox reactions. Even if organic memristive devices cannot be considered as ideal memristors, in which memristance depends on the magnetic field flux, but more likely they are threshold-type memristive elements, the flux linkage could indirectly provide precious information on switching conditions, selecting accurately the integration interval.

OMDs voltages thresholds depend not only on some of the characterization parameters, such as scan speed, but also on sample properties, such as the thickness. In our case, since the oxidation threshold is 0.4 V, to calculate the flux corresponding to the ON switching of the device (i.e. the PANI oxidation process), we selected the integration range based on the Gaussian fit of the positive peak in the gate circuit. Following this definition (Eq. 6.3):

$$\varphi = \int_{t_{0.2V}}^{t_{0.7V}} V_{SD} dt \qquad (6.3)$$

and reporting the calculated value versus the noise amplitude, it is possible to appreciate a rather perfect agreement between the intensity of the negative peaks in Fig. 6.25c and the flux values reported in the inset of the figure. This suggests that this anomalous variation in the ΔI is qualitatively correlated in intensity to the flux calculated in the specific interval.

Even varying the noise administration scheme, applying a signal in which v_{noise} is dependent on the instantaneous memristance of the device, the resultant effect is coherent with what was previously reported. Since the major effect of the perturbation is confined in the voltage ranges in which the organic memristive device is in a high resistivity state, hysteresis curves (Fig. 6.26a) present the highest perturbation in the OFF state based on gain parameter (Fig. 6.26c).

Even in this case, the hysteresis is present (Fig. 6.26a) for all the explored gain values but it resulted shifted to lower voltage values. This effect is confirmed by the presence of two negative peaks in the corresponding dependences for ΔI (Fig. 6.26b). This effect finds, as in the previous case, a qualitative description in the flux linkage calculation performed in the interval range between $t_{0.2V}$ and $t_{0.7V}$ (Fig. 6.26d).

In summary, the investigation of the effects induced by the application of noised signals on the performance of organic memristive devices has demonstrated both a significant modification of devices hysteretic response and an increase in the probability of resistance switching at lower voltage values. These effects find a good correlation with the flux linkage calculated during the resistance switching process. To quantify and easily depict these variations, we introduced a new performance parameter for organic memristive devices that is the representation of the shift

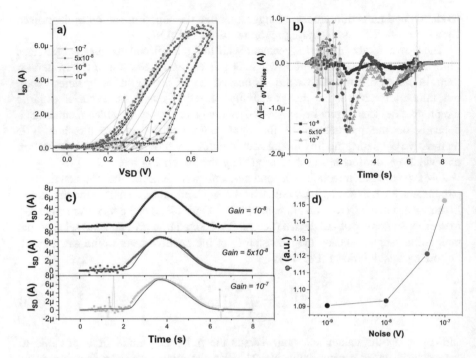

Fig. 6.26 Organic memristive device: response to noise sourcing: (**a**) I-V hysteresis behavior as a function of different noise amplitudes; (**b**) ΔI vs Time (in this case ΔI is calculated using as reference curve the one obtained with gain $= 10^{-9}$); (**c**) Temporal evolution of I_{SD} for different noise amplitudes plotted with respect to the reference curve obtained with gain $= 10^{-9}$ and (**d**) Flux for the oxidation process vs noise amplitude. (Reprinted from Battistoni et al. [357], Copyright (2020), with permission from Elsevier)

between different cycles of the same measurements (ΔI). This parameter well describes the aging of organic memristive device performances as well as the effect of the noised sourcing.

6.7 Frequency Driven Short-Term Memory and Long-Term Potentiation

In this section, we will consider the effects of the appearance of short-term memory and long-term potentiation (learning) in organic memristive devices, making parallels with synaptic plasticity in the nervous systems of living beings.

The initial temporal diagram for the voltage and current variations that will be used in this section is shown in Fig. 6.27.

As it is shown in Fig. 6.27, the value of the output current of the organic memristive device is determined by the value of the applied voltage, which is higher (lower) than the oxidation (reduction) potential and time interval when it was

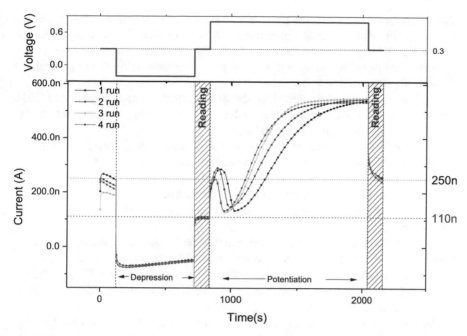

Fig. 6.27 Temporal dependence of the output current of an organic memristive device as a function of the applied voltage: upper panel: temporal profile of the applied voltage; lower panel: output current of the organic memristive device. The readout was done applying the voltage of +0.3 V, depression of the conductivity was done by applying −0.2 V, potentiation of the conductivity was done by applying +0.8 V [358]. (Reprinted from Battistoni et al. [358], Copyright (2019), with permission from Elsevier)

applied. One can control the resistance of the device by varying one of these parameters.

The stabilized value of the current remains constant after the reinforcement or suppression phases for all cycles. The kinetics of the potentiation and depression are different, as was explained in Chap. 2. In this study, we are particularly interested in the position of the oxidation peak during the potentiation phase (corresponds to the 800–100 seconds in Fig. 6.27) that shifts continuously to left during successive cycles. This behavior is directly due to the successive transfer of different zones of polyaniline in the active area, and it can be connected to the Hebbian rule in its initial presentation [7]: *When an axon of cell A is near enough to excite cell B and repeatedly or persistently takes part in firing it, some growth process or metabolic change takes place in one or both cells such that A's efficiency, as one of the cells firing B, is increased.*

This property of the organic memristive device is suitable for neuromorphic applications [156, 306, 358] and the realization of model systems, mimicking human memory [359, 360].

In the model of Atkinson and Shiffrin [360], the incoming information after a very short-time memorizing at the level of sensor memory is transferred from the short-term memory level to the permanent long-term memory due to the repeatable processes with the probability proportional to the number of successful repetition cycles. Results on mimicking such processes with memristive devices are presented in [359]. The switching of operation modes was determined by the application of pulses with different durations.

Similar results were obtained using the organic electrochemical transistor [361], where the switching between short- and long-term memory was a result of the amplitude of the signal, applied to its gate.

In order to check such transitions in our organic memristive devices [358], initially, +0.8 V and −0.2 V were applied to the device for 10 minutes. As it is clear from Fig. 6.28a, the variation of the current value related to its initial value was about 600% in the case of potentiation and about 86% in the case of depression. These values determine limits of the effectiveness of the training using pulses as input stimuli.

It was studied the influence of the number and polarity of the applied high-frequency stimuli (pulses of 20 ms) on the value of the output current. Figure 6.28b, c show the voltage pulses used for the potentiation and depression. Color points in Fig. 6.28d, e show experimentally measured values of the relative current variation on the number of pulses shown in Fig. 6.28b, c.

As is clear from Fig. 6.28d, a small number of positive pulses does not vary significantly the conductivity of the device with respect to its initial state. Application of each pulse results in a small increase of the current that relaxes to the initial value after the end of the pulse. Such behavior observed also previously [359, 361] can be considered as short-term potentiation (temporal reinforcement of synaptic weight) that relaxes to the initial state when the action of stimulus is finished [359, 360, 362]. As the number of the applied pulses increases, we can see a gradual increase of the conductivity till the saturation (400%) that occurs after 1500 passed pulses. Thus, in the case of the use of high-frequency pulses, the organic memristive device demonstrates the transfer from the short-term memory mode to the long-term potentiation mode, depending on the number of passed pulses, which is in good agreement with processes in biological samples.

In the case of negative pulses application, we can see the suppression of the conductivity (red squares in Fig. 6.28e). However, short-term suppression was not observed in this case. Even a small number of passed pulses (100 during 2 seconds) resulted in a significant decrease of the output current value (about 10%).

The results shown in Fig. 6.28 demonstrate the possibility of the transfer from short-term to long-term memory in systems, based on organic memristive devices, which is a fundamental step for the realization of neuromorphic functions, such as long-term potentiation and depression of connections, depending on the activity of elements.

The questions about the role of pulse parameters are widely discussed by research groups working in the field of memristive devices. However, in the case of inorganic

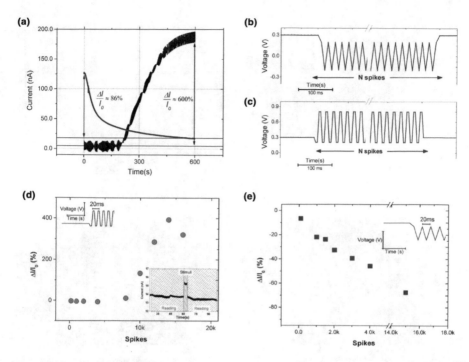

Fig. 6.28 Long-term potentiation and suppression of conductivity using various voltage pulses: (**a**) temporal variations of the value of the output current when applying positive (black curve) and negative (red curve) voltages; (**b**) voltage pulses used to suppress conductivity; (**c**) voltage pulses used to enhance conductivity; (**d**) the dependence of the change in the relative value of the output current on the number of applied pulses in the amplification node (top insert – the used pulse profile; bottom inset demonstrates that the application of limited number of positive pulses does not affect the conductivity of the device); (**e**) the dependence of the change in the relative value of the output current on the number of applied pulses in the suppresion mode (in the box, pulse profile is presented) [358]. (Reprinted from Battistoni et al. [358], Copyright (2019), with permission from Elsevier)

memristive devices, the results were obtained using pulses with different amplitudes [71, 363, 364]. In the biological samples, instead, these processes are connected to the frequency and number of the applied stimuli. Moreover, the shape of the pulse generated by neurons implies the presence of two picks of opposite polarity corresponding to depolarization and hyperpolarization of the membrane, as is shown in Fig. 6.29a by the red line.

The synaptic weight can be potentiated or depressed depending on the frequency of stimuli that arrived at the nerve cell. Few seconds of giant stimuli (high frequency of signal arriving) results in the potentiation of the synaptic connection, while long-time application of stimuli with low frequency of their repetition results in the

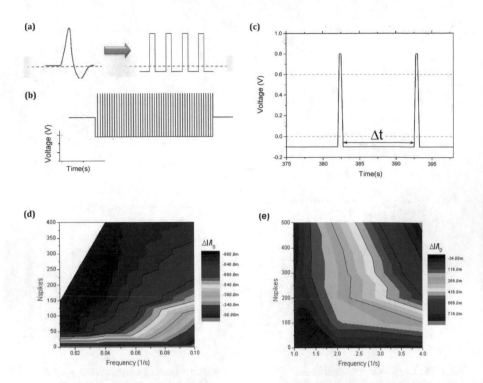

Fig. 6.29 Long-term potentiation and depression of conduction with pulses of different frequencies: (**a**) red curve – a form of the potential in biological systems, the blue curve corresponds to the pulses applied in these series of experiments, green dotted line – "the potential of relaxation," applied between the pulses; (**b**) time dependence of the amplitude of the applied potential in the experiment (the value of the conductivity was measured on the read-phase (red line) before or after learning (blue section)); (**c**) an enlarged view of a pulse sequence used for the learning process (dotted lines correspond to the oxidation (green) and reduction (yellow) potential); (**d**) color chart of relative changes of output current depending on the number of past pulses and frequencies in the regime of long-term suppression of conductivity; (**e**) color chart relative change of the output current depending on the number of past pulses and frequencies in the regime of long-term enhancement of the conductivity [358]. (Reprinted from Battistoni et al. [358], Copyright (2019), with permission from Elsevier)

depression of this connection [363]. In [365] it was suggested to use memristive devices with diffusive dynamics for mimicking such behavior. In this case, the device was switched to the high conducting state if the applied voltage was higher than a certain threshold value. When the voltage was switched off, the device relaxed to the insulating state. However, the difficulty of the relaxation prediction in devices of such type results in the practical impossibility of reaching stable intermediate levels of the conductivity, different from those in the most conductive and most insulating states. In this respect, the organic memristive device has a very important advantage: there are several intermediate states of the conductivity corresponding to the ratio of oxidized and reduced fractions of polyaniline in the active zone.

Characteristics of organic memristive devices have been measured in the potentiation mode, varying the frequency of the applied pulses and duration of each pulse. The amplitude of pulses was constant. Pulses of 0.5 s and 0.8 V were used in this experiment. A constant offset of -0.2 V was used for realizing a pulse shape, corresponding better to the shape of the signal in biological systems. The frequency is related to the time between successive pulses (Eq. 6.4):

$$f_i = 1/\Delta t_i \qquad (6.4)$$

Frequencies used in this experiment were: 1.0, 1.33, 2.0 and 4.0 Hz (high frequency branch) and 0.1, 0.04, 0.02 and 0.01 Hz (low frequency branch). The conductivity state of the organic memristive device after passing a certain number of pulses at a defined frequency was measured by applying +0.3 V, as was described above.

The results are shown in Fig. 6.29.

Color diagrams showing the relative variations of device current as a function of passed pulses in the case of low and high frequencies are shown in Fig. 6.29d, e, respectively. Application of stimuli with low frequency results in the depression of the conductivity (Fig. 6.29d): the lower the frequency is, the fewer pulses are required for the device switching into the insulating state. In the case of higher frequencies (Fig. 6.29e), we can see the increase in the device conductivity. Reaching the max conductivity state demands a simultaneous increase of the number of passed pulses and their frequency. When one of these parameters is decreased, we can see a significant decrease in the conductivity potentiation.

In the case of the depression of the conductivity, the low-frequency range is more effective. The lower the frequency is, the fewer pulses are required for the conductivity suppression.

Thus, diagrams shown in Fig. 6.29d, e are in good agreement with short-term potentiation and depression observed in biological objects [364, 366]. Moreover, they provide an additional demonstration of the transfer from short-term to long-term potentiation.

6.8   Spike-Timing-Dependent Plasticity (STDP) Learning in Memristive Systems

Real neuromorphic information processing must provide also the possibility of unsupervised learning. Currently, the main paradigm of unsupervised learning is connected to spike-timing-dependent plasticity (STDP) mechanism [367, 368]. This mechanism is responsible for the automatic induction of causal low, and it is responsible for the variation (reinforcement or inhibition) of synaptic connections according to the time delay between spiking of pre- and postsynaptic neurons. If we consider a spike at nervous cell 1 (presynaptic) as an event 1 and the spike at the nervous cell 2 (post-synaptic) as an event 2, the STDP mechanism is responsible for making correlations between these events. When spikes of Event 1 and Event 2 occur with long temporal delay, these events are not connected, and the synaptic weight does not change. Instead, when the time interval between events is short (we are speaking about millisecond scale), the synaptic weight can be varied. If Event 1 occurs before Event 2, the synaptic weight is reinforced (as shorter is the delay, as stronger is the increase of the weight function): it means that the nervous system attribute automatically the role of the cause to Event 1 and the role of the consequence to Event 2. If, instead, Event 2 happens before Event 1, the weight function of the synaptic connection is inhibited, meaning that Event 1 cannot be the reason for Event 2. Importantly, as shorter is the delay between these events, as stronger is the variation of the synaptic connection between these nerve cells.

Currently spiking neuron networks are considered as the most prospective systems for the realization of circuits capable of unsupervised learning [369–372] due to the possibility of realizing casual connections, similar to that we have in the nervous system.

As memristive devices are considered as synapse analogs in electronic circuits, several systems with spiking learning algorithms, similar to STDP, were realized and tested [373–375].

STDP-like learning has been also demonstrated with organic memristive devices [376]. It is to note that practically ideal STDP dependences have been obtained with such devices [299]. Of course, the timescale was larger than in biological samples, but it can be adjusted by scaling the devices to smaller sizes [377].

SDTP-like algorithms were applied for the realization of classical conditioning (Pavlov dog learning) on electronic circuits with organic memristive devices, based on polyaniline [299] and parylene [211, 305] materials. One question can arise: is the case of Pavlov dog supervised or unsupervised learning? The presence of the experimentalist points out the supervised nature (where the STDP mechanism also works). However, the association of the food with sound was done by the dog itself. Therefore, the question is rather a philosophic one and it is not so easy to distinguish supervised and unsupervised contributions during this learning process.

Organized arrays of organic neuromorphic devices, immersed in an electrolyte, revealed also characteristics similar to the homeoplasticity phenomena of the neural environment [378].

In this section, we will consider STDP-like learning algorithms for systems based on two types of organic memristive devices: polyaniline-based (main object of this book) and parylene-based. After discussing basic principles of realized circuits, we will consider their application for the realization of systems, allowing classic conditioning, as it is in the case of the Pavlov dog.

6.8.1 STDP in Circuits with Polyaniline-Based Memristive Devices

For the STDP implementation, the connected gate and drain electrodes were assigned as a presynaptic input, and the source electrode was considered as a postsynaptic one [376]. Identical potential pulses were used as pre- and postsynaptic spikes, but a constant bias voltage of +0.2 V was applied to avoid changes in the device conductivity between spikes conditioned by the value of the PANI redox equilibrium potential. Amplitudes of the spikes were chosen to be 0.3 V so the maximum potential difference across a memristive element was equal to +0.8 V, and the minimum one was equal to −0.4 V. Temporal profiles of the pre- and postsynaptic pulses are shown in Fig. 6.30a. Postsynaptic pulses were applied after presynaptic pulses with a delay time Δt, which can have both positive and negative values. Actually, the interchange of pre- and postsynaptic electrodes varies only the sign of the delay of the Δt values. The form of resulting pulses and the definition of Δt are shown in Fig. 6.30a. Temporal variation of the total voltage in the system (the difference between post- and presynaptic potentials) across the memristive element during the measurement for the delay time $\Delta t = 200$ s is shown in Fig. 6.30b.

Conductances were measured by the application of the testing voltage of +0.3 V within 30 s before and after the pre- and postsynaptic pulses sequences. Generally, in

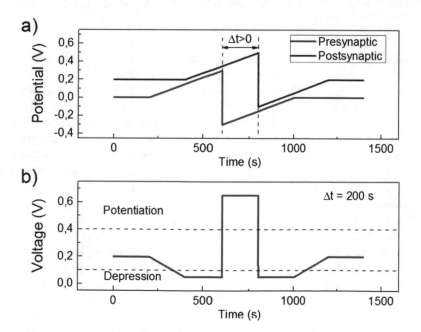

Fig. 6.30 (a) Shapes of presynaptic (red line) and postsynaptic (black line) potential pulses. (b) Resulting voltage across the memristive element for the specific $\Delta t = 200$ s value [376]. (Reprinted from Lapkin [376], Copyright (2018), with permission from Elsevier)

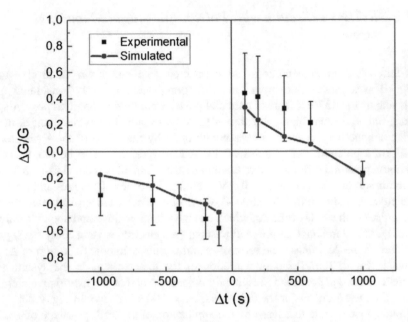

Fig. 6.31 STDP window for the organic PANI-based memristive element – relative weight (conductance) changes for different delay Δt values: experimentally obtained (black squares) and simulated (red circles and lines) [376]. (Reprinted from Lapkin [376], Copyright (2018), with permission from Elsevier)

neuromorphic applications of memristive elements, synaptic weights are equal to their conductance. Thus, weight changes were considered as $(G_{fin} - G_{ini})/G_{ini}$, where G_{fin} and G_{ini} are the final and initial conductance values respectively. We measured weight changes due to the STDP for $\Delta t = \pm 1000, \pm 600, \pm 400, \pm 200$, and ± 100 s, each time resetting conductance value before the application of spikes. Thus, determined weight change dependence on delay time (STDP window), averaged over several samples, is shown in Fig. 6.31. On the base of the phenomenological memristive element kinetic model, we have simulated the application of the same voltage pulses for the selected delay times and the result is also shown in Fig. 6.31.

The variation of weights decreases monotonically with the increasing delay time and a transition from negative to positive values occurs at the delay time of 0 s. The observed decrease of the weights for long time delay values Δt, both negative and positive can be explained by the difference of the time of the switching between conducting and insulating states of organic memristive devices (switching from conducting to insulating state is much faster than the switching from insulating to conducting state, as was discussed in Chap. 2). The same reason is responsible for the fact that we see fewer weight changes for the positive Δt values than that for the negative ones. The results of the simulation demonstrate the qualitatively similar behavior and the quantitative estimations of the conductance change are within the limits of an experimental error (Fig. 6.31). It should be also noted that the conductance retention time of the PANI-based memristive devices under equilibrium potential is about 24 hours. It is not enough for practical use, but it could be elongated (i) instrumentally, by periodical monitoring and correction of resistive state, (ii) synthetically, by optimization of the composition of polymer materials used in this work, e.g. by adding metallic nanoparticles in PANI layer [379].

Of course, from the practical point of view, the characteristic switching times of PANI-based memristors should be reduced. There are some promising efforts on the way to reach this goal such as a decrease of the active PANI film thickness down to a few molecular layers and optimization of chemical and electrical properties of electrolyte used in a memristive element.

In fact, the results were significantly improved when the thickness and lateral dimensions of the PANI channel were reduced [299]. The following results were obtained for samples, when the distance between the source and drain electrodes d was 10 or 20 μm, and its thickness was 6 or 10 PANI molecular layers, transferred from the water surface.

The main idea of the STDP implementation in electronic circuits with memristive devices is that the pre- and postsynaptic pulse amplitudes are somehow summed up. If the time delay between these pulses is less than the total pulse duration, the pulses overlap changing the voltage on the memristive device. Such a resulting pulse may exceed the oxidation threshold and, as a result, the conductance of the device will increase. In contrast, if the resulting pulse amplitude falls below the reduction potential, the device conductance will decrease. The relative change in conductance (in %) depends on the time that the device was held above or below the corresponding threshold (oxidation or reduction) potential. In turn, this time depends on the delay Δt between the pulses.

In this experiment, bipolar rectangular voltage pulses were used, since they allow to obtain more stable and reproducible STDP windows [299, 380]. In the first half of the pulse, the voltage was kept at the level of +0.2 V for 5 s in the second half at −0.2 V for 5 s. These parameters were selected so that a single voltage pulse does not significantly affect the device conductance. Such pulses were applied to the pre- and postsynaptic electrodes with the time delay of Δt.

A constant bias of +0.4 V was applied to the postsynaptic electrode to prevent changes in the conductance in time intervals between spikes. The conductance values G_{before} and G_{after} were measured at this voltage value before and after the spike sequences were applied, respectively. After that, the relative conductance variation was calculated according to the following formula (Eq. 6.5):

$$\Delta G/G = |G_{after} - G_{before}|/G_{before} \tag{6.5}$$

To avoid an accumulation of the conductance change from one measurement to the other, the memristive device was brought to the initial state. It was potentiated under the voltage of +0.8 V for 20 s for the negative delays and depressed under the voltage of −0.3 V for 0.2 s for the positive ones. The obtained STDP window (the

Fig. 6.32 STDP window for the memristive device—relative conductance changes for different delay Δt values. The inset shows the shapes of presynaptic (black) and postsynaptic (red) potential pulses [299]. (Reproduced with permission from Prudnikov et al. [299]. Copyright (2020) IOP Publishing, Ltd)

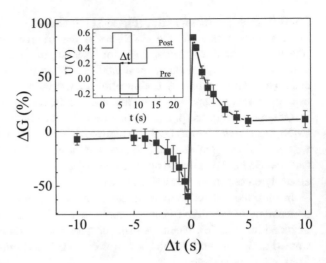

dependence of the relative conductance change $\Delta G/G$ on the time delay Δt between pre- and postsynaptic pulses) is shown in Fig. 6.32.

The dependence presented in Fig. 6.32 resembles very well a typical STDP window observed in biological systems [367]. The conductance change is negative ($\Delta G < 0$) for negative delays $\Delta t < 0$ and, vice versa, positive ($\Delta G > 0$) for positive delays $\Delta t > 0$ promoting a causal relationship. Temporarily separated spikes (the time between the spikes longer than the single spike duration) do not affect the conductance as well as a single spike does not.

In comparison to the STDP window measured for macroscopic PANI-based devices [376], the microscopic devices allow the reduction of the characteristic times by two orders of magnitude. Even if these times are still more than those observed in biological systems, they are more applicable for the realization of artificial spiking neural networks.

6.8.2   STDP in Circuits with Parylene-Based Memristive Devices

In this section, we will consider memristive devices based on another organic compound – parylene. The repetitive unit of this polymer is shown in Fig. 6.33.

This material is widely used in the field of organic electronics [381–385], including memristive devices [116, 386–388]. One of the most promising memristive structures for "wearable" applications are Metal-Insulator-Metal structures based on polymeric layers of parylene (poly-para-xylylene, or PPX) due to the simple and cheap production of this polymer, its transparency, and the possibility of films fabrication on flexible substrates [116, 206]. Moreover, parylene is a US Food

Fig. 6.33 Parylene
Repetitive Unit

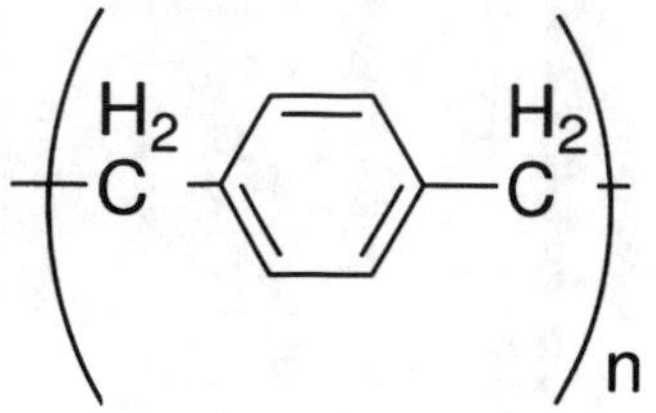

and Drug Administration approved material and could be used in biomedicine since it is completely safe for the human body, which cannot be said about most of the other organic materials [206–208]. Currently, parylene-based structures have shown good memristive characteristics [206] including their capability of multilevel resistance switching.

Parylene-based memristive devices were not considered yet in detail in this book. As results obtained on devices fabricated with this material seem important, we will make a short overview in this section. In particular, we will consider the mechanism responsible for the resistance switching in parylene-based metal-insulator-metal structures.

The element in this study has the following structure [305]: metal/parylene/indium tin oxide (M/PPX/ITO structures). The parylene layers (~100 nm) were deposited on ITO coated glass substrates (bottom electrode, or BE) by the gas phase surface polymerization method. At the used vacuum levels, all sides of the substrate were uniformly impinged on by the gaseous monomer, resulting in a truly conformal coating. The ITO-coated glass was selected as the BE because it has wide commercial availability and benefits such as high conductivity, transparency, and resistance to moisture.

The top metal electrodes (TEs) were fabricated from Ag, Al, or Cu layers (~500 nm thick) obtained by thermal evaporation or ion-beam sputtering through shadow masks. The sizes of the TEs were 0.2×0.5 mm^2, and about 150 single devices (per one substrate) were fabricated for each kind of electrode. The metals listed above were selected due to their widespread use in electronic engineering including the fabrication of memristive devices.

M/PPX/ITO memristive devices have generally good performance, especially when Cu is used as top electrode. Indeed, the Cu/PPX/ITO structures (Cu samples) demonstrated the R_{off}/R_{on} value of ~10^3, endurance higher than 10^3 cycles, and at least 16 stable resistive states with retention time >10^4 s for $R_{on} = 1$ kΩ and $R_{off} = 1$ MΩ.

Figure 6.34a shows several successive I–V cycles measured on one Cu sample, along with their averaged I–V curves, which represents the so-called cycle-to-cycle (C2C) stability. The measurements were carried out at room temperature under normal conditions, the external voltage was applied to the Cu TE with the ITO BE grounded. Current compliance of 1 mA was adopted to prevent a device breakdown. As one can see, the Cu/PPX/ITO structure practically does not need a forming process and every sweep goes in a rather similar way. The device-to-device (D2D)

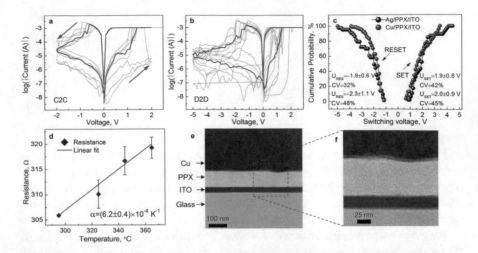

Fig. 6.34 Electrophysical and structural characterization of the M/PPX/ITO structures. (**a**) I–V cyclic characteristics showing the typical bipolar resistance switching behavior of the Cu/PPX/ITO sample during seven cycles (cycle-to-cycle variability); the average curve is highlighted in bold. (**b**) I–V characteristics were collected in eight different Cu/PPX/ITO devices (device-to-device variability, the fifth of 10 cycles is shown for each); the averaged characteristics are highlighted in bold. (**c**) Cumulative probabilities of USET and URESET switching voltages and their coefficients of variation (CV) for ~100 I–V cyclic characteristics measured in the samples with copper (red) and silver (black) top electrodes. (**d**) Temperature dependence of the low resistance state resistance of the Cu/PPX/ITO structure. (**e**) Cross-sectional TEM image of the Cu/PPX/ITO sandwich structure. (**f**) Enlarged image of the area highlighted by the rectangle in (**e**), showing roughness of the Cu/PPX interface [305]. (Reprinted from Minnekhanov et al. [305], Copyright (2019), with permission from Elsevier)

curves along with their average characteristics are presented in Fig. 6.34b. It is clear that the cycles for different samples have reasonable repeatability, which is also confirmed by the distribution of resistance switching voltages U_{SET} and U_{RESET} (Fig. 6.34c). This distribution was found to be narrower (the coefficient of variation is lower) for the Cu samples compared to the Ag ones. Considering the symmetrical and repeatable behavior of the resistance switching, we can suggest that it is most likely caused by the formation of a conducting filament in the dielectric film between the electrodes. In our case, it is reasonable to assume that such a conducting filament is a metal bridge consisting of atoms of the top electrode (formed as a result of electrochemical metallization) [389, 390].

The temperature-dependent switching characteristics of the aforementioned Cu/PPX/ITO memristive devices were studied, as they are informative for understanding the resistance switching mechanisms. The resistance of the structure was calculated from the I–V measurements conducted in the range from 0 to 0.5 V. Figure 6.34d shows the dependence of the low resistance state resistance (R_{on}) on temperature. It is clear that this resistance increases linearly with temperature, exhibiting the typical metallic conduction property. In general, the dependence of metallic resistance on temperature can be expressed as $R(T) = R_0[1 + \alpha(T - T_0)]$, where R_0 is the resistance at T_0 and α is the temperature coefficient of resistance. According to the experimental data (Fig. 6.34d), we have obtained the coefficient $\alpha = (0.62 \pm 0.04) \times 10^{-3}\,\mathrm{K}^{-1}$ at 300 K. It should be noted that the α of Cu nanowire decreases with its diameter, probably due to surface diffuse scattering [391, 392]; thus, we may conclude that the metallic behavior of the LRS of Cu/PPX/ITO devices originates from the small-size conducting Cu filament.

The TEM investigations also show the possibility of the formation of metallic bridges between the top and the bottom electrodes of the M/PPX/ITO structures. Indeed, one can see in Fig. 6.34e–f that the copper layer is not completely smooth and the roughness of the Cu/PPX interface is sufficient for the metal ions to begin the migration toward the cathode (ITO).

The process of resistance switching in the M/PPX/ITO structures could be explained by the electrochemical mechanism: the metal ions of the top electrode move into the polymer layer under the action of positive voltage, then migrate to the bottom electrode reducing on it, and form a conducting filament connecting the top and bottom electrodes (see Fig. 6.35a–d). When a negative voltage is applied (Fig. 6.35e–f), the thinnest part of the filament ruptures due to the Joule heating some of the metal ions return to the top electrode and the structure switches to the high resistance state.

Samples with Cu electrodes were chosen for STDP learning experiments due to the reasons listed above. The bottom electrode (ITO) of the Cu/PPX/ITO memristive structure was assigned for the presynaptic input, and the top electrode (Cu) was considered as the postsynaptic one. Identical voltage pulses were used as pre- and postsynaptic spikes of heteropolar bi-rectangular (inset in Fig. 6.36a) or bi-triangular (inset in Fig. 6.36b) shape. The amplitudes of birectangular and bi-triangular spikes were 0.7 V and 0.8 V, respectively. Therefore, the spike itself could not lead to a conductivity variation in the structure. On the other hand, if two spikes are summed up, the potential difference across the memristive device could be increased up to ± 1.4 V and ± 1.6 V, which is within the resistance switching range of the Cu-containing sample. The pulse half-durations were 150 and 200 ms with the discretization of 50 ms. Postsynaptic pulses were applied after (before) presynaptic pulses with varying delay time Δt (ranged from -500 to 500 ms with a step of 50 ms).

Conductance was measured by applying a testing voltage of +0.1 V for 50 ms before and after the sequence of pre- and postsynaptic pulses. Generally, the device conductance G is regarded as a synaptic weight, and then its change (ΔG) is equal to a synaptic weight variation. More specifically, a weight variation corresponds to $\Delta G = G_f - G_i$, where G_f and G_i are the final and the initial conductance values, respectively. Thus, the determined weight variation dependences on the delay time are shown in Fig. 6.36.

As it is clear from Fig. 6.36, the experimental results have demonstrated the rule similar to STDP one observed in biological systems [212]. Synaptic potentiation ($\Delta G > 0$) was observed for $\Delta t > 0$, and synaptic depression ($\Delta G < 0$) was observed for $\Delta t < 0$. Note that the result of STDP-like learning depends on the G_i value. If a

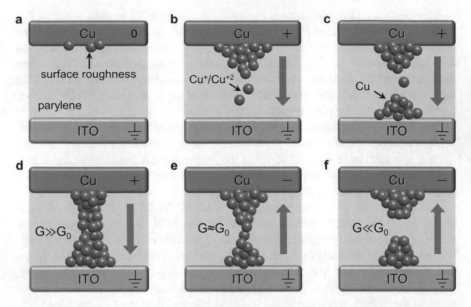

Fig. 6.35 Schematic representation of the evolution of metal bridges (conducting filaments) in Cu/PPX/ITO memristive devices and the consequent quantum conductance effect. (**a**) Fragment of a pristine sandwich structure, having some surface irregularities on the top electrode. The orange pellets represent Cu atoms. (**b**) A positive voltage is applied to the top electrode of the structure; copper ions begin to move to the cathode (ITO) under the action of an electric field. (**c**) Copper ions reach the bottom electrode and reduce, so a conductive filament begins to grow. (**d**) The conductive filament was completely formed; quantized conductance is not observed. (**e**) A negative voltage is applied to the top electrode; copper ions begin to move backward to it. A quasi-point contact is formed, so the conductance is quantized, becoming approximately equal to G_0. (**f**) The conductive filament has ruptured; conductance is much less than G_0 [305]. (Reprinted from Minnekhanov et al. [305], Copyright (2019), with permission from Elsevier)

memristor state is close to low resistance one, then its synaptic weight would likely depress rather than potentiate (as for the 1 mS state in Fig. 6.36), and vice versa (0.1 mS state in Fig. 6.36). On the other hand, when the memristor initially was in the intermediate state (0.5 mS), learning curves demonstrated both synaptic potentiation (up to 120% for $\Delta t > 0$) and depression (down to −44% for $\Delta t < 0$). This "multiplicative" character of the memristive STDP curve can be explained by taking into account the finiteness of a conductance variation in the studied memristors.

It should be noted that a wide range of pulse amplitudes (0.5, 0.6, 0.7, 0.8, and 0.9 V) and half-durations (50, 100, 150, 200, and 250 ms) were studied for several resistive states of the Cu sample. Results were similar to the reported ones.

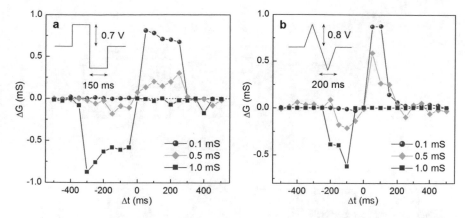

Fig. 6.36 STDP window of Cu/PPX/ITO memristive structures (for various initial conductance values) obtained with heteropolar (**a**) bi-rectangular and (**b**) bi-triangular spike pulses shown in the figure insets. Post-synaptic spikes were applied after (before) presynaptic ones with a varying delay time Δt. Every point of the curves is an average of 10 recorded experimental values [211]. (Reproduced from Minnekhanov et al. [211])

6.8.3 Classic Conditioning of Polyaniline-Based Memristive Devices Systems

To demonstrate the possibility of the memristive devices' application in neuromorphic systems capable of STDP learning, a model circuit mimicking Pavlov's dog behavior has been developed [299]. It consists of two inputs ("food" and "bell") and one output.

As this behavior represents classic conditioning, several circuits based on memristive devices have been realized and reported [155, 301–304]. Similar results have been obtained also with polyaniline-based memristive devices in DC [300] and pulse [242] modes, even if they were not directly connected to the Pavlov dog learning mimicking. However, the STDP-like algorithm was applied only in [299].

A triangular-shaped pulse was chosen as inputs due to the two main advantages. Firstly, it is more energy-efficient. Secondly, biological neurons are supposed to propagate signals with bi-triangular spikes, which means that our pulses are also more bio-plausible.

Although forward conditioning was used in this experiment, the backward one can also be achieved using these devices but was not directly possible with the scheme implemented in the reported research. This type of learning requires the anti-STDP rule, which can be obtained by changing either the polarity of the memristive device or the voltage pulses [393].

Following the idea of the conditioning, the output current before training should be high when the following combinations of the stimuli are activated: only "food" and both "food" and "bell" simultaneously. Low output is expected when only the "bell" input is activated before training. Training should result in the fact that the

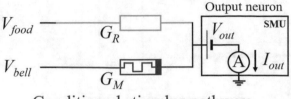

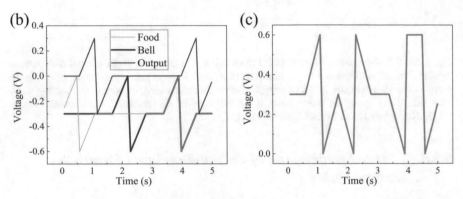

Fig. 6.37 (**a**) Scheme of Pavlov's dog model implementation using bi-triangular pulses allowing training of the system by the STDP-like mechanism. The unconditioned pathway is implemented with a constant resistor and the conditioned pathway—with a memristive element; (**b**) training voltage pulse sequence repetitively applied to the inputs and output. (**c**) Resulting potential difference across the memristive device [299]. (Reproduced with permission from Prudnikov et al. [299]. Copyright (2020) IOP Publishing, Ltd)

"bell" input stimuli must lead to a high output current, which means that salivation is caused. In addition, the "food" input activation must also lead to high output, which should not be significantly changed since this output represents an unconditioned response.

The "food" input is connected to a resistor, and the "bell" input to an input of the PANI memristive device, while both outputs of the resistor and the memristive device are connected to an output neuron. A sequence of voltage pulses is applied to the inputs. At each time step, the neuron reads the current and if the current exceeds a threshold value it generates a voltage pulse at the next time step. The time step duration was 80 ms. The scheme of the system is shown in Fig. 6.37a.

A pulse sequence used in these experiments is shown in Fig. 6.37b. At the beginning of the training sequence, the "food" input pulse was applied. The "food" input activation causes "high" output current (higher than the neuron threshold value) and thus triggers an "output" pulse. Thus, the "food" input pulse is accompanied by the output pulse, as is shown in Fig. 6.37b (first 1.5 s). The "bell" input activation before the training is followed by a "low" output current due to the low-conductive state of the memristive device. Thus, the output pulse is not gener- ated after the "bell" input pulse. When both the "food" and "bell" inputs are activated simultaneously, the "output" pulse caused by the "food" input interferes with the pulse from the "bell" input and the resulting voltage on the memristive device forms a plateau with an amplitude of +0.6 V (Fig. 6.37c) that is enough for increasing the memristive device conductance. The spike parameters (baseline and amplitude) were selected to avoid a noticeable variation of the conductance by a single pulse. This pulse sequence was applied to the circuit repetitively and no inputs were discon- nected during the experiment. Thus, after several simultaneous activations of both the "food" and "bell" inputs, the memristive device conductance becomes high enough (higher than the neuron threshold) for the output pulse being triggered by the "bell" input pulse only. Note that learning, in this case, is possible only for the specific timing between the output and input pulses: if the time difference between them is positive, indicating the causal relationship, then the potentiation of the memristor could be achieved, which is in a good agreement with the STDP learning principles.

The threshold for the output neuron necessary for the generation of the output pulse was chosen to be 30 μA that corresponds to a single "food" input stimulus- response before the training procedure. After a few cycles, the "bell" response increased significantly and exceeded a 30 μA threshold, as it is shown in Fig. 6.38a. It means that the "bell" pulse itself causes "salivation" by producing a high output current.

To reveal the effect of the voltage pulse duration, training pulses of various durations of 320, 400, and 480 ms (half-length) were used; all other parameters were the same. As is clear from Fig. 6.38b, if the training pulse duration is increased, the number of cycles required to achieve the desired "high" output for the "bell" input is decreased. The total required training time remains approximately the same for all the pulse durations, as it is shown in Fig. 6.38c. Therefore, the main factor affecting the training speed is the time when the memristive device is under the voltage exceeding the potentiating threshold of +0.5 V.

Figure 6.38a has also one interesting feature: the output current, corresponding to the only "food" tends to increase slightly (green). It can be explained taking into account that there is a current also in the chain of the memristive device. Since the device is not disconnected, a constant bias voltage is applied, causing the additional current even when the "bell" input is inactive. The growing conductance of the memristive device during the training causes the increase of the total current.

6.8.4   Classic Conditioning of Parylene-Based Memristive Devices Systems

After successful experimental implementation of STDP-like learning for parylene- based memristive structures, a step forward to demonstrate their utility in

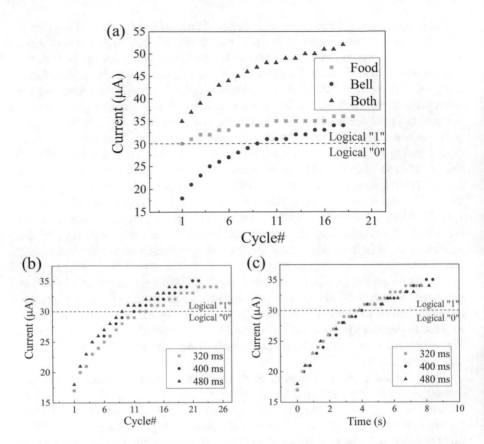

Fig. 6.38 An output current variation during repetitive applications of the input voltage pulses for (**a**) "bell" (red circle), "food" (green square), and both (blue triangle) stimuli and (**b**), (**c**) only the "bell" stimulus with different durations plotted versus (**b**) the number of cycles and (**c**) full time of training in a case of triangular-shaped pulses. The dashed red line represents a selected threshold current. (Reproduced with permission from Prudnikov et al. [299]. Copyright (2020) IOP Publishing, Ltd.)

constructing simple neuromorphic networks has been done. For this purpose, the task of classical (also known as Pavlovian) conditioning was chosen [212, 213]. The constructed network (simulating Pavlov dog behavior) consisted of 2 presynaptic neurons connected with a postsynaptic one (Fig. 6.39a). The first presynaptic neuron corresponding to an unconditioned stimulus (e.g. "food") pathway is connected to the postsynaptic one via a resistor R. The second presynaptic neuron connection is represented by a memristive element (Cu/PPX/ITO) corresponding to an initially neutral stimulus (e.g. a "bell") pathway. Each neuron was implemented in software: the presynaptic neurons were programmed to generate spikes of amplitude U_{sp}, and the postsynaptic one was used as a threshold unit (generating spikes only in the case of the total input current exceeding the threshold current value I_{th}, which is chosen to

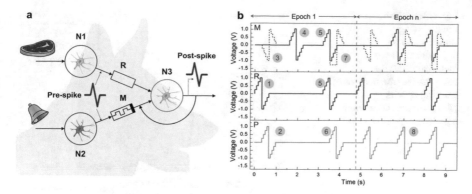

Fig. 6.39 STDP-like learning memristive Pavlov dog implementation. (**a**) The electrical schematic diagram: N1 — the 1st pre-neuron, spiking after the "food"-related stimulus; N2 — the 2nd pre-neuron, spiking after the "bell" stimulus; N3 — the post-neuron, which spikes when the total input current exceeds the threshold; R — a resistor with a constant resistance value of $R = 2 \text{ k}\Omega$; M — a memristive element, initially in the $R_{off} = 20 \text{ k}\Omega$ resistive state. A post-spike is generated unconditionally after a spike comes from N1 and under the condition that the memristor current exceeds I_{th} after a spike comes from N2. (**b**) An example of the spike pattern applied to the inputs of the scheme: 1 — the initial pulse (1st Epoch) on the resistor (R) (unconditioned stimulus), resulting in post-spike (P) 2, which in turn comes to the memristive device (M) as pulse 3 (dashed) in the inverted form; 4 — the pulse on the memristive device, initially without post-neuron activity; 5 — simultaneous pulses on the resistor and the memristive device, which result in post-spike 6 leading to the training pulse 7 (dashed); 8 — a post- spike as a result of the conditioned stimulus when the training is completed (Epoch n, where n is equal to or above the number of epochs required for successful conditioning) [211]. (Reproduced from Minnekhanov et al. [211])

be slightly less than the ratio U_{sp}/R). The bottom electrode of the memristive element was connected to the output of the postsynaptic neuron. A similar electronic implementation of the Pavlov dog has been presented before [155], however, there was used constant-signal learning without the use of any STDP-like rules. Another implementation was proposed in a network with pseudo-memcapacitive synapses with a Hebbian-like learning mechanism [215].

The learning procedure was as follows: (1) introducing a signal only down the unconditioned stimulus pathway (by this step we check the correct post-synaptic neuron activity, i.e. the dog starts "salivating" after it has been exposed to the sight (or smell) of "food"); (2) sending a signal only down the conditioned stimulus pathway (in this step we check whether an initially neutral stimulus becomes a conditioned one); (3) pairing the two stimuli (in this step the conditioning (learning) occurs). These three steps constituted one epoch of learning shown schematically in Fig. 6.39b.

Resistance $R = 2 \text{ k}\Omega$ (Fig. 6.39a) was chosen to be slightly higher than the R_{on} resistive state of the Cu/PPX/ITO memristive structure (~1 $k\Omega$) to provide the possibility of successful training of the network. The spike amplitudes U_{sp} and durations Δt_{sp} were identical for all neurons and were selected experimentally according to the results of the I-V and STDP-like learning measurements

Fig. 6.40 Electronic
Pavlov's dog
implementation:
dependence of the
memristive device
resistance on the number of
learning epochs
[211]. (Reproduced from
Minnekhanov et al. [211])

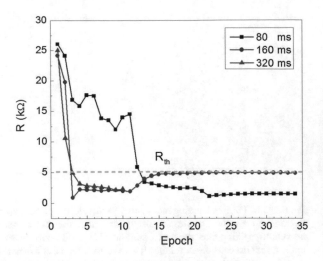

[211]. Every experiment started with the $R_{off} \sim 20$ kΩ state of the memristive device
(which was, in fact, an intermediate state, because resistance in the "true" high
resistance state reaches 5 MΩ). As it is shown in Fig. 6.39b, when the conditioned
stimulus is paired with the unconditioned one (Fig. 6.39b, spikes 5), the postsynaptic
pulse (spike 6) from N3 (which starts being generated when the memristive device
current exceeds the threshold value I_{th}) sums up with the presynaptic pulse from N2,
leading to a resistivity change in the memristor (by the dashed spike 7). If the
introduction of the "bell" signal alone resulted in the postsynaptic neuron spiking
(spike 8), the conditioning was successful.

In the experiments, the parameters were such that the threshold resistance value of
the memristive device required for the neuron spiking (i.e. the highest resistance
value when $U_{sp}/R_{th} > I_{th}$), was $R_{th} \sim 5$ kΩ. Spikes of heteropolar bi-triangular shape
with the amplitude of 1 V and the half-duration of 80, 160, and 320 ms were used. It
can be seen from Fig. 6.40 that the conditioning occurred in all the cases, but the
number of epochs necessary for conditioning was different depending on the pulse
durations. Namely, the shorter was the pulse duration, the larger was the number of
epochs necessary for successful conditioning. This may be due to the fact that the
short duration of the applied electric field (80 ms) is not enough for the formation of
continuous metal bridges. Since the effect of local heating can be neglected here
(a relatively long time passes between the pulses — about 1 s), the total pulse time
remains the only parameter determining the conditioning rate (assuming constant
amplitude of learning pulses). The obtained results demonstrate the associative
learning ability of neuromorphic systems with parylene-based memristive elements
and provide a basis for the development of autonomous circuits capable of emulating
some cognitive functions.

6.9   Coupling with Living Beings

An important application of memristive devices could be connected to their coupling with living beings. In fact, as these devices have some features similar to the elements of the nervous system, several possibilities of such coupling have been discussed in the literature [394, 395]. Critical analysis of similar features of natural and artificial neural networks can be found in [396, 397].

On the other hand, several biological materials have revealed memristive properties when they were used as active materials of such devices. Among these materials, there are works on plants [398–401], slime mold [402–409], silkworm [410], and even human skin [411].

However, the coupling of living beings with memristive devices was not the subject of the mentioned works. Authors used natural biological materials as active media for the realization of elements with memristive properties.

In the case of inorganic memristive devices, it has been reported a coupling of artificial spiking neuron networks with embryonic hippocampal rat nerve cells. Modulation of the transmission strength in the biological medium was induced by the plasticity of memristive devices [412].

In a complimentary interesting study on organic neuromorphic devices, the authors were able to mimic modulation of the strength of connections according to the presence of neurotransmitters. The device was based on a three-terminal device with PEDOT:PSS active layers. Rat pheochromocytoma cells were deposited on the gate of this device, and the conductivity of the channel was modulated by the dopamine presence demonstrating long-term conditioning and synaptic weights recovery in a flow chamber [413].

An important step (even considering possible synapse prosthesis applications) has been done when two live nerve cells from the rat cortex were coupled through the organic memristive device [414]. The results were obtained with polyaniline-based devices (the main subject of this book) and present, probably, the only available information on the direct experimental coupling of neuronal cells of the brain with memristive devices. Therefore, we will consider this work in more detail.

A synapse is a biological structure, which connects two neurons enabling specific and unidirectional information flow (excitation or inhibition) from one neuron to another. Synaptic connections are the key elements of the neuron networks and their plasticity underlies learning and memory. Recent progress in building artificial neuronal networks is largely based on the elements mimicking features of natural synapses [415–417]. The use of electrophysiological spike-sorting [418] and optogenetic [419] approaches enable an efficient readout and control over the activity of single or groups of neurons that can lead to the development of prosthetic devices. Restoration of synaptic connections, as in the case of traumatic injury, as well as in other pathologies associated with a synaptic loss of function and various synaptopathy could be also solved through an introduction of electronic synapses to connect neurons directly, given that these artificial synapses recapitulate the main feature of natural synapses including their plasticity. Moreover, the development of

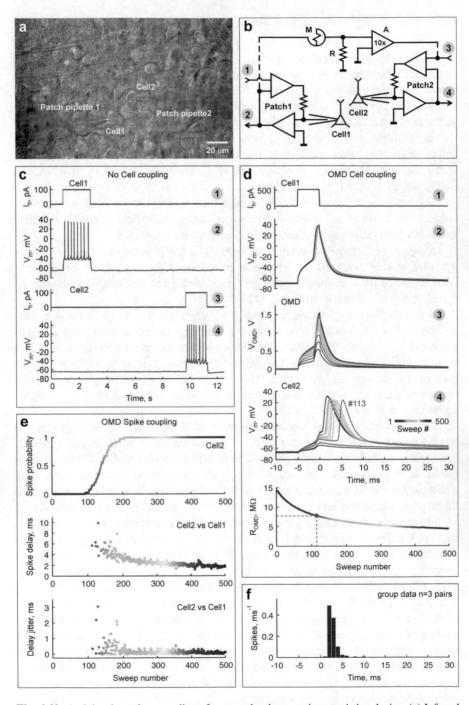

Fig. 6.41 Activity-dependent coupling of neurons by the organic memristive device. (**a**) Infrared differential interference contrast microphotograph of a P7 rat brain slice with visually identified L5/6 neocortical cells (Cell1,2) recorded simultaneously. (**b**) Simplified electrical scheme of two patch-clamp amplifier headstages (Patch1,2): 1,3—patch-clamp holding inputs; 2,4—patch-clamp

electronic synapses with unprecedented, due to biological restraints, features during evolution could result in the creation of cyborgs with unprecedented capacities.

In the described experiments, in order to be protected from different artifacts, patch-clamp recordings from nonconnected pairs of cortical layer five pyramidal neurons in rat brain slices (Fig. 6.41a) were used. Action potentials (APs) evoked by suprathreshold depolarizing current injection in either neuron failed to evoke any response in another cell in the pair (Fig. 6.41c), indicating that these cells were not connected by natural synapses in either direction. These neurons were then connected through an electronic circuit with an organic memristive device playing the role of a synapse analog (Fig. 6.41b). After setting the organic memristive device resistance initially at high values by negative voltage loading (Fig. 6.41d, bottom), APs in a "presynaptic" Cell 1 (Fig. 6.41d, plot 2) were induced by suprathreshold depolarizing step (Fig. 6.41d, plot 1). However, these APs in Cell 1 evoked only a subthreshold depolarizing response in the "postsynaptic" Cell 2 (Fig. 6.41d, plot 4, the first blue color-coded sweeps) due to the high initial organic memristive device resistance (Fig. 6.41d, bottom panel, and Fig. 6.41e, top panel). Since the resistance of the organic memristive device reduces upon depolarization [194], the consecutive depolarizing steps and the APs in Cell 1 induced a gradual increase in voltage responses from the organic memristive devices (Fig. 6.41d, plot 3) and Cell 2 (Fig. 6.41d, plot 4). When the resistance of the organic memristive device was decreased by a factor of ≈2 (Fig. 6.41d, bottom, sweep #113), the depolarizing response in Cell 2 reached the AP threshold (≈−40 mV) and Cell 2 started reliably to fire APs (Fig. 6.41d, plot 4, sweep #113 onwards and Fig. 6.41e). Along with a further decrease in memristive device resistance (Fig. 6.41d, bottom), the firing probability of Cell 2 gradually increased (Fig. 6.41e, top). The activity-dependent increase in spike coupling between neurons was also associated with an improvement in the spike-timing through the organic memristive device synapse as evidenced by a progressive reduction in the AP delays (Fig. 6.41d, plot 4, and Fig. 6.41e, middle plot), and a reduction in the jitter of AP delays (Fig. 6.41d, plot 4, and Fig. 6.41e, bottom plot) [361, 420]. Figure 6.41f shows the histogram

---

Fig. 6.41 (continued) primary outputs; and an organic memristive device-based circuit (5 × 5 mm) connecting two neurons. (**c, d**) Traces of current-clamp recordings from Cells 1 and 2 before (**c**) and after (**d**) coupling through the organic memristive device. Traces 1–4 correspond to the inputs/outputs as labeled in b. Note that prior to coupling through the organic memristive device (**c**), APs in either neuron failed to evoke responses in the other neuron, indicating that these cells were not connected by natural synapses. After the connection of Cells 1 and 2 through the organic memristive device (**d**), the efficacy of coupling progressively increases with each consecutive depolarizing step/AP in Cell1. 500 traces (color coded by sweep #) are aligned with suprathreshold depolarizing steps delivered to Cell 1. The bottom plot, organic memristive device resistance as a function of the sweep #. Dashed lines indicate the first sweep when Cell2 started firing. (**e**) Corresponding plots of the activity-dependent change in spike probability in Cell2 (top), spike delay of Cell2 from Cell1 (middle), and spike delay jitter in Cell2 (bottom). (**f**) Histogram of the spike delay in Cell2 from Cell1 calculated for three OMD-coupled cell pairs (777 spikes) [414]. (Reprinted from Juzekaeva et al. [414], John Wiley and Sons, © 2018 WILEY-VCH Verlag GmbH & Co. KGaA, Weinheim)

summarizing the delay times of Cell2 spiking from t Cell1 for three cell pairs (777 spikes recorded). It is noteworthy that the characteristic timing of the AP commutation through the OMD-synapse is similar to that of natural excitatory synapses [421].

To further assess the synaptic response, we examined if coupling with organic memristive devices could enable also neuronal synchronization during spontaneous activity. To this end, presynaptic Cell 1 was continuously depolarized by injection of constant inward current to allow spontaneous firing (Fig. 6.42a, blue trace). We observed that the firing of Cell 1 induced a gradual decrease in resistance of the organic memristive device (Fig. 6.42d, bottom), which in turn induces an increase in Cell 2 responses as in the experiments above (Fig. 6.42a, red trace). As soon as the suprathreshold level of cell coupling was achieved, Cell 2 started firing APs in synchrony with Cell 1 (Fig. 6.42a, d), yet with a 3.8 ± 0.1 ms time lag ($n = 3$ cell pairs, 633 spikes recorded; Fig. 6.42e). The synchronized firing of the neurons coupled through the organic memristive device occurred in the δ-frequency range (0.56 ± 0.04 Hz, $n = 3$ cell pairs; Fig. 6.42b, c) that is characteristic of the slow-wave cortical activity during deep sleep [422].

In this study, it has been provided for the first time the experimental evidence of the unidirectional, activity-dependent coupling of live neurons through the organic memristive device. It has been demonstrated that the spike-timing features of the artificial synapses based on organic memristive devices approach those of the natural excitatory synapses, that the magnitude of this coupling can be controlled by the neuronal activity, and that these artificial synapses efficiently support neuronal synchronization in a simple two-neuron network.

Electronic devices (organic and inorganic) have been used to record extracellular field changes [423], cellular action potential [424–426], and in vivo electrophysiological recordings [427]. Despite the large interest, the realization of a prototype of artificial synapse chips is mostly focused on the controlled release of chemical compounds [428] through micropatterned substrates [429, 430] more than on the creation of a functional interface between devices and cells. Such kind of connection must allow the signal transmission between cells and device and, most importantly, the transmission of stimulations between groups of cells through the device. There are several works where silicon circuits were used for mimicking synapse functions [415, 416]. Several memristor-based circuits have demonstrated plasticity suitable for artificial neural networks [283, 431], mimicked the learning of simple animals [155, 156], and been used for the acquisition and partial decoding of signals from the retina [257]. However, the evidence that memristive devices could functionally couple live neurons was still missing. This is the primary requirement for their implementation in prosthetic devices or in building hybrid networks that requires the use of special materials in adequate configurations. Furthermore, if we consider the perspective of implantable systems, one must fulfill requirements such as biocompatibility, flexibility, and stretchability. In this framework, here we demonstrate a quite novel functionality with the first evidence of unidirectional, activity-dependent coupling of two live neurons in brain slices via organic memristive devices. The coupling is characterized by nonlinear relationships determined by the instantaneous

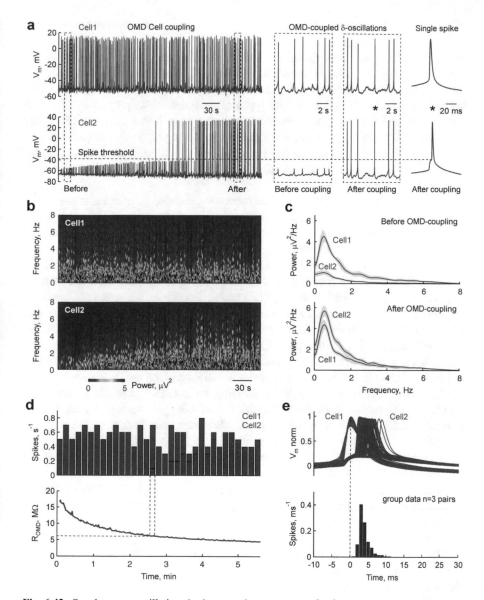

Fig. 6.42 Synchronous oscillations in the natural neuron network where two cortex neurons were coupled through the organic memristive device. (**a**) Current-clamp recordings from Cell1 (blue) and Cell2 (red). Parts of traces outlined by dashed boxes before and after spike coupling through the organic memristive device are shown on expanded time scales on the right. The horizontal dashed line indicates Cell 2 spike threshold. (**b**) Corresponding membrane potential spectrograms in Cells 1 and 2. (**c**) Power spectrum density plots of the membrane potential in Cells 1 and 2 before (top) and after (bottom) spike-coupling through the organic memristive device (confidence interval is shadowed; $n = 3$ pairs). (**d**) Frequency of spikes (top) in Cell 1 (blue) and Cell 2 (red) calculated for the 10 s bin intervals from the recordings shown in a and the corresponding values of the organic memristive device resistance (bottom). Dashed lines indicate the onset of spike-coupling between Cells 1 and 2. (**e**) Example of 65 normalized spikes (top) recorded in Cell 1 (blue traces) and Cell 2 (red traces) and the histogram of the spike delay in Cell 2 from Cell 1 (bottom), data from three organic memristive device-coupled cell pairs (633 spikes) are pooled together [414]. (Reprinted from Juzekaeva et al. [414], John Wiley and Sons, © 2018 WILEY-VCH Verlag GmbH & Co. KGaA, Weinheim)

resistance of organic memristive devices, depending on the connected neuronal activity, and by the observed excitation threshold in the postsynaptic neuron. We have demonstrated that the coupling through organic memristive devices shows also spike-timing features similar to those types of natural excitatory synapses. Furthermore, our artificial synapses, based on organic memristive devices, efficiently support synchronized delta-oscillations in the two-neuron network. All such features ever observed simultaneously in such a simple single device circuit make organic memristive device-based synapses eligible for a significant step toward the realization of prosthetic synapses.

For considering these devices as elements for synapse prosthesis, some major important properties are required: (1) they must show plasticity in signal transmission emulating that typical in living beings; (2) for implants, devices must be flexible and made of biocompatible materials; (3) size of the elements must be comparable with cell dimensions; (4) interconnections and external electronics system should not significantly disturb the functioning of the nervous system. Each of these features requires appropriate studies and optimizations. For items (2) and (3), there are already positive solutions. In fact, since the early works, it has been shown that organic memristive devices can be realized in a flexible configuration [306]. As far as the biocompatibility of the materials for organic memristive devices is concerned, it has been demonstrated in experiments of cell growth [432, 433] and simple organisms [158, 163, 434] on their surface. Moreover, the active channel can be fabricated by the LbL technique [253], allowing the formation of architectures where all the parts in contact with the biological surrounding are made of biocompatible polymers [255, 435]. Requirement 3 has also been demonstrated, showing the scalability of the organic memristive devices [233] by fabricating devices with lateral sizes of 20 μm [377] that can be scaled further, as it has been demonstrated the possibility of the realization of submicron organic electronic systems [436].

Thus, the major challenges concern properties (1) and (4) that require further studies and developments. In particular, the question of interconnects and external electronics (4) is very critical in the case of brain implants, a subject of large efforts by several groups [415, 416, 437].

The aim of this work was to address and contribute to developing the major requirement (1), hence concentrating our efforts to demonstrating for the first time directly that organic memristive devices can have the same plasticity of electrical properties of the chemical synapses in living beings.

The major outcome of this study is that we provide experimental evidence of unidirectional, activity-dependent coupling of live neurons through an organic memristive device. We have demonstrated that the spike-timing features of artificial synapses based on organic memristive devices are very similar down to several details to those of natural excitatory synapses, that the magnitude of coupling through organic memristive devices can be regulated by neuronal activity, and that these synapses efficiently support neuronal synchronization in a simple two-neuron network. We, therefore, give important indications that organic memristive devices, apart from being key candidate elements for neuromorphic computational systems (where up to now purely electronic elements have been used for information storage and processing [194]), also should be considered as suitable elements for developing "synapse prosthesis" and useful for neuromorphic computational systems, where the same elements will be used for memorizing and processing of the information.

Experimental details of these experiments can be found in [414].

Realization of synapse prosthesis in the brain is a very difficult task, as the number of synapses is very large, and they are organized in 3D structures. Synapse prostheses in other parts of the nervous system seem to be an easier task. For example, it is possible to consider the repairing of locomotor activity making prosthesis of damaged parts of the spinal nervous system [438, 439].

To make a short summary, in this chapter, we have considered several neuromorphic applications where it has been demonstrated that organic memristive devices have several important properties of biological synapses that allow to use them in artificial electronic circuits, mimicking the learning processes of living beings. In addition, it was also shown the possibility of coupling living nervous cells through organic memristive devices, which open several perspectives towards the realization of synapse prosthesis.

Chapter 7
3D Systems with Stochastic Architecture

In Chap. 6, we have demonstrated that organic memristive devices can be effectively used as synapse analogs in artificial electronic circuits, mimicking some functions of the nervous system. However, shown examples required the use of max two memristive devices. Design of systems, allowing complex information processing, learning, and decision-making require using much more elements. For example, it is estimated that the human brain contains 10^{14}–10^{15} synapses. Moreover, they are organized in 3D systems with short and long connections between neurons. The direct approach for the realization of such high level of integration is the use of modern lithography methods. Despite the fact that significant progress has been achieved in this direction, most realized systems have currently a planar organization. There are several works, demonstrating the possibility of the realization of multilayer inorganic structures, but the number of layers in them is very limited [71, 440–444]. Instead, the brain nerve cells are organized in 3D systems, allowing multiple pathways for signals.

In this respect, it seems very perspective to consider the so-called "bottom-up" approach, based on the capability of some organic molecules to form self-assembled complicated 3D structures [445–459]. It is to note that this approach is absolutely not applicable to inorganic systems, where lithography methods form still the main-stream up to now.

For the neuromorphic systems, based on organic memristive devices, such systems must contain three main components of the device: conducting polymer, electrolyte, and insulator. Mutual orientations and connections of these materials can have a stochastic character. If the number of junctions will be high, the possibility of the formation of multiple signal pathways will appear.

In this chapter, we will consider three types of such 3D systems with a stochastic organization: free-standing fibrillar systems, layered systems on skeleton supports, and systems based on phase separation of materials.

7.1   Free-Standing Fibrillar Systems

This system was the first realized one [450], and it is based on the stochastic crossing of polyethylene oxide and polyaniline free-standing fibers.

The basic idea of this approach was the following one: the presence of the fibrillar system between input and output electrodes will provide the formation of multiple possible signal pathways. According to the Hebbian rule [7], the conductivity of an individual pathway, composed of many synapse-like junctions, will be increased with the time (or frequency) of its involvement in signal propagation. There is also the possibility of its suppression after the application of adequate external action, as it takes place in the case of supervised learning.

The fibrillar systems were fabricated in the following way.

As the first step, a fibrillar network of polyethylene oxide was realized. For this reason, the drop of polyethylene oxide (0.1–0.5 ml) water solution with 0.1 M LiCl and 0.1 M HCl was placed onto the glass support with evaporated electrodes. A silver wire (usually, 50 μ diameter) was placed in the middle of the drop. This sample was placed into the vacuum chamber and pumped to the pressure of 10^{-2} Torr. After it, the pumping was continued for additional 15–20 minutes. This treatment has resulted in the formation of a fibrillar polyethylene oxide structure on the glass support. After it, the drop (0.1–0.2 ml) of polyaniline solution was placed onto the already formed sample and pumped during 15–20 minutes after reaching the pressure of 10^{-2} Torr.

Images of different samples, acquired with an optical microscope, are shown in Fig. 7.1.

Figure 7.1 demonstrates the formation of the mono-component (a) and bi-component (b) fibrillar polymeric system. Fibers of both materials have a diameter of about several tens of microns and up to several centimeters in length. Despite the stochastic and unpredictable character of the distribution of these fibers, the presence of numerous intersections allows to expect the numerous possible signal pathways, the conductance of which can be controlled electrochemically, as it was for individual organic memristive devices.

In order to check the validity of this suggestion, structures with two metal electrodes on the glass support and a silver reference electrode in the middle of the fibrillar system, placed in the drop before the vacuum treatment, were investigated. It was expected that if the stochastic distribution of fibers provides the required Ag – polyethylene oxide – polyaniline configuration, we must see cyclic voltage-current characteristics, similar to those of organic memristive devices.

Typical experimentally measured cyclic voltage-current characteristics of such samples are shown in Fig. 7.2.

The characteristics shown in Fig. 7.2 have a rectifying behavior and small hysteresis, typical for organic memristive devices, indicating the formation of expected heterojunctions. Qualitative similarity was observed for practically all measured samples: the shape was the same, but the values of the current were different from sample to sample due to the stochastic origin of formed structures.

Fig. 7.1 Images of
samples, composed from
polyethylene oxide (**a**) and
polyethylene oxide –
polyaniline (**b**), acquired
with an optical microscope
(image sizes: 0.6 × 0.5 mm)
[450] (Reproduced from
Erokhin et al. [450] with
permission from The Royal
Society of Chemistry)

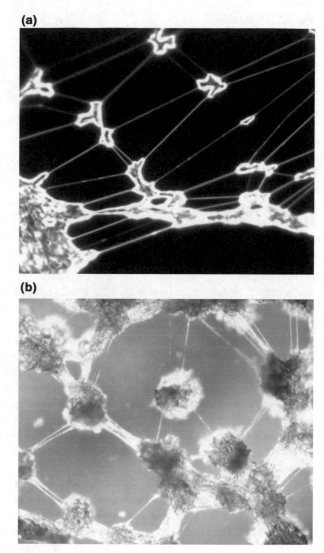

The next step was to check the possibility of the system adaptation, as was done on the eight-element circuit, described in Chap. 6.

The system contained one input and two output electrodes. The training was performed for the realization of high signal at one output electrode and low one at the other, applying stimulus to the input electrode. The value of the current in the output electrodes circuit, when +0.4 V was applied to the input electrode, was considered as the output. As we discussed in previous chapters, this voltage value should not vary the conductivity state. Therefore, current values at output electrodes will indicate the conductivity of each circuit. The training was done by applying different potentials

Fig. 7.2 Cyclic voltage-current characteristics of the 3D fibrillar system of polyethylene oxide and polyaniline [450]. (Reproduced from Erokhin et al. [450] with permission from The Royal Society of Chemistry)

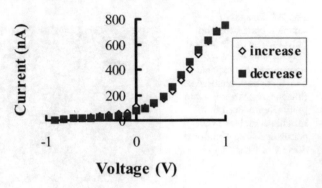

Table 7.1 Training of the fibrillar stochastic network

	In-Out1 (nA)	In-Out2 (nA)
Before training	20	20
After training	200	20

to the output electrodes. In particular, +1.2 V was applied to the first electrode, while −0.6 V was applied to the second one. Results of this system state before and after training are summarized in Table 7.1.

Table 7.1 demonstrates the possibility of the potentiation of the chosen signal pathway, similar to that, reported for the deterministic network, composed from eight discrete devices, in Chap. 6. The importance of this result must be underlined: the network does not have a defined architecture with fixed connections of elements, but it was composed of stochastic connections between fibers of conducting and ionic polymers. However, the complexity of the network with multiple possible signal pathways allowed to potentiate a conductance between the defined pair of electrodes and depress it between the other one by applying the adequate training procedure.

However, this system has one very important drawback – very low stability of its structure and properties. The variation of the reached conductivity state was observed practically from the beginning of measurements. After 40 minutes of measurements, the conductivity was decreased for one order of magnitude, and the system stops operations after 2 hours. This result cannot be considered as a surprising one taking into account the free-standing nature of the formed network. Current, passing in fibers, results in their Joule heating and, as a result, the appearance of temperature gradients, leading to the deformation and even complete destruction of the fibrillar structure.

Therefore, for having the possibility of considering this system for practical applications, it must be stabilized in some way. The utilization of stabilizing porous frames seems to be a perspective way, because even nature uses rigid skeletons for stabilizing the organization of many living beings. Such attempt has been done [306], and the image of the realized structure, obtained with an electron microscope, is shown in Fig. 7.3.

Fig. 7.3 Fibrillar system, formed on a porous frame. (Reprinted from Erokhin et al. [306], Copyright (2010), with permission from Elsevier)

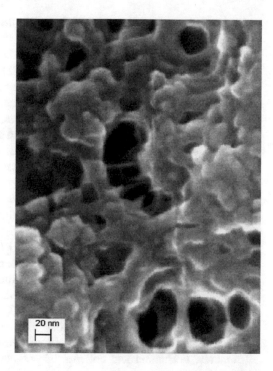

As it is clear from Fig. 7.3, this approach allows to diminish the fiber diameter up to 10 nm, which, in principle, can lead to a significant increase of the integration level of the network and speed of its operation. Such formed structures are even more stable, maintaining their structure even during imaging (combined action of a rather high vacuum and electron beam irradiation).

Despite the fact that positive results have been obtained in this direction and the first demonstration of the possibility of training of stochastic network was done on these systems, improvement of stability and characteristics demanded the search for other solutions, two of which will be considered in the successive sections.

7.2   Stochastic Networks on Frames with Developed Structure

There are a lot of objects with a stable developed structure in nature. In this section, we will consider a stochastic adaptive network, realized using a sponge as a stabilizing frame. Obviously, the main technique, used for the realization of conducting channels, namely, the Langmuir-Blodgett technique, discussed in Chap. 2, will be not applicable for these supports: it does not allow to cover all surfaces of the sponge. Therefore, it was studied the possibility to form conducting channels with a method that will allow to cover all surfaces through the whole sponge depth.

As it was considered in the previous subsection, stochastic adaptive networks can be based on polyethylene oxide-polyaniline fibrillar networks. However, the free-standing nature of this system results in its very low stability. In order to avoid the degradation of the structure and properties, the method that can be often seen in nature has been chosen. Stabilization of systems, based on soft matter (polymers, in our case), demands the use of skeletons – ridged and stable frames. In the case of the realization of 3D systems, these frames must have developed porous structures, allowing the penetration of materials, necessary for the realization of organic memristive devices.

A natural cellulose sponge was chosen to be used as this frame. The sample had the following sizes: 15 mm length, 8 mm width, and 5 mm height. Pore sizes were within 0.1–0.5 mm [435].

Initially, the sponge was immersed into a 0.1 M solution of hydrochloric acid for 10 hours. After it, the sample was dried on a filter paper and immersed into the polyaniline solution for 30 min. After it, the sample was again dried on filter paper. Silver wire was inserted into the central part of the sample. Polyethylene oxide water solution (20 mg/ml), containing 0.1 M of $LiClO_4$ and 0.1 M HCl, was injected with a syringe (the volume of about 0.1 ml) into the central part of the sample. The sample then was placed into an exicator and vacuumed with a mechanical pump for 30 minutes. Formation and collapse of polyethylene oxide bubbles were clearly visible on the sample surface. After this procedure, three gold electrodes (wires) were mechanically connected to the sample, as is shown in Fig. 7.4.

One of the gold electrodes was grounded, while two others were connected to independent power supplies through amperemeters, measuring current values in these circuits. A silver electrode was also grounded through another amperemeter, allowing to measure the value of the ionic current in the system.

The measurements were done in a closed chamber, where a glass with a 0.1 M solution of HCl was placed.

Before making training experiments, cyclic voltage-current characteristics were measured in both possible signal pathways: S-D1 and S-D2. The measurements of the variation of the ionic current were done in parallel. These characteristics are shown in Fig. 7.5.

The characteristics, shown in Fig. 7.5, demonstrate the success of the realization of multiple polyaniline-polyethylene oxide junctions, revealing two important properties of organic memristive device: hysteresis and rectification. The other important feature of the realized system is that the value of the ionic current is one order of magnitude less than the value of the total current. It demonstrates the optimal composition of the electrolyte, used in these experiments: the amount of ions is enough for redox reactions and effective resistance switching, but the concentration of these ions is low enough for not contributing significantly to the value of the total conductivity (stochastic nature of the sample does not allow to decrease further this concentration).

Fig. 7.4 Schematic representation of the sample, used for the training of the stochastic 3D system, assembled on the porous frame (**a**); image of the sponge, covered by polyaniline (sizes: 15 mm × 10 mm × 5 mm) (**b**) [435]. (Reproduced from Erokhina et al. [435] used in accordance with the Creative Commons Attribution (CC BY) license)

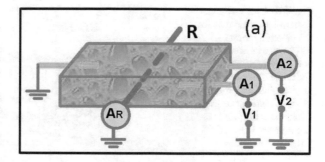

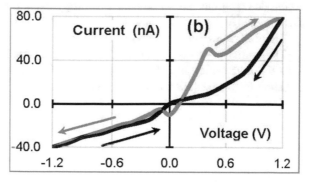

Fig. 7.5 Cyclic voltage-current characteristics, measured between S and D electrodes, shown in Fig. 7.4 for total (**a**) and ionic (**b**) currents. Current values were acquired 60 seconds after the potential application. Arrows indicate the direction of the variation of the applied voltage [435]. (Reproduced from Erokhina et al. [435] used in accordance with the Creative Commons Attribution (CC BY) license)

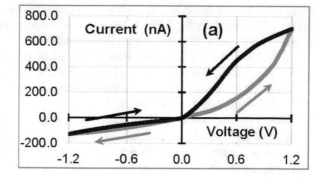

Fig. 7.6 Cyclic voltage-current characteristics for the electron current (difference between the total and ionic currents). Arrows show the direction of the applied voltage variation [435]. (Reproduced from Erokhina et al. [435] used in accordance with the Creative Commons Attribution (CC BY) license)

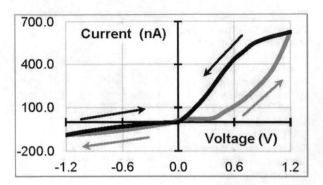

Table 7.2 Training of stochastic system, assembled on porous frame (Reproduced from Erokhina et al. [435] used in accordance with the Creative Commons Attribution (CC BY) license)

	S-D1 (nA)	S-D2 (nA)
Before training	50	55
After training	200	21

Cyclic voltage-current characteristics for the differential current, demonstrating the electrons component of the total conductivity, is shown in Fig. 7.6.

The comparison of characteristics, shown in Figs. 7.5 and 7.6 with those for fibrillar networks, presented in the previous subsection, shows that in the present case we have not only the rectification, but also more pronounced hysteresis, which is the first indication that this system is more stable.

After the demonstration that both branches of the network have memristive characteristics, experiments on the training of the whole system were done.

The training implied the potentiation of the conductivity in one branch (S-D1) and its depression in the other one (S-D2).

Similarly to the experiments on fibrillar networks, three values of the voltage were used: two training and one testing voltages. The value of the testing voltage was chosen in such a way that its application does not vary the conductivity state. In this case, it was +0.3 V. Training voltages must vary the conductivity state. Thus, S electrode was grounded, while +1.0 V was applied to D1 and −0.2 V was applied to D2 simultaneously for 60 minutes.

Training results are presented in Table 7.2.

The results, presented in Table 7.2, demonstrate that in this case, we can have not only the potentiation but also suppression of the signal pathways, contrary to the situation for free-standing fibrillar networks.

7.3   3D Stochastic Networks, Based on Phase Separation of Materials

In this section, we will consider questions connected to the realization and investigation of a stochastic 3D network of memristive devices using a process of phase separation of block copolymers, containing parts of solid electrolyte (polyethylene oxide) and insulator (polystyrene sulfonic acid). In addition, this system contains polyaniline (material with variable conductivity) and gold nanoparticles. The use of gold nanoparticles is determined by the following reason. Memristive devices represent well synapse properties. However, natural and artificial neuron networks demand also the presence of threshold elements (neuron bodies), allowing the income of all possible signals but providing the output only if the sum of input signals during a certain time interval is higher than some defined threshold value. As the conducting pathways of the network must be fabricated from polyaniline, it was suggested that the distribution of gold nanoparticles can play the role of such threshold elements. In fact, significant difference in work functions of gold and polyaniline will result in the possibility of signals to enter the particle, but the exit will be possible only when the Schottky barrier will be overcome [451, 452].

The complex approach, used in this section, has demanded a successive study of all necessary components and technological steps.

7.3.1   Stabilized Gold Nanoparticles

As was mentioned above, functionalized gold nanoparticles [453] are an essential part of the system. These particles have been synthesized according to the available methods [454–456]. Schematic representations of the used gold nanoparticles are shown in Fig. 7.7.

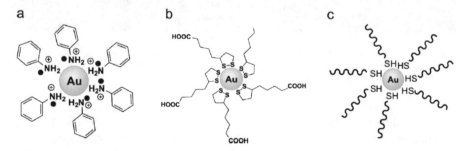

Fig. 7.7 Schematic representations of gold nanoparticles with the surface functionalized by aniline (**a**), lipoic acid (**b**), dodecane, and octane thiol (**c**) (difference is only in the hydrocarbon chain length) [453]. (Reprinted from Berzina et al. [453], Copyright (2011), with permission from Elsevier)

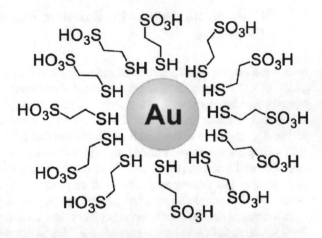

Fig. 7.8 Gold nanoparticles, stabilized by 2-mercaptoethanol sulfonic acid [453]. (Reprinted from Berzina et al. [453], Copyright (2011), with permission from Elsevier)

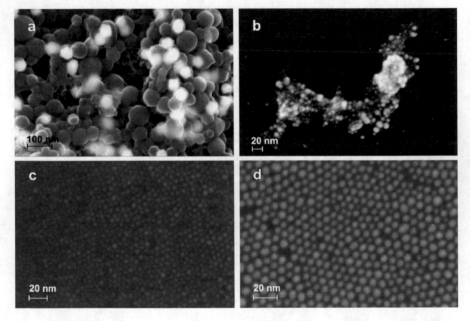

Fig. 7.9 SEM images of arrays of gold nanoparticles, stabilized by aniline (**a**), lipoic acid (**b**), dodecane thiol (**c**), and octane thiol (**d**) [453]. (Reprinted from Berzina et al. [453], Copyright (2011), with permission from Elsevier)

In addition, nanoparticles, whose surface was functionalized by 2-mercaptoethanol sulfonic acid, were also used. These particles were synthesized according to methods described in [250, 251, 457], and their schematic representation is shown in Fig. 7.8.

SEM images of samples, prepared by drying of cast solutions of nanoparticles, shown in Fig. 7.7, are presented in Fig. 7.9.

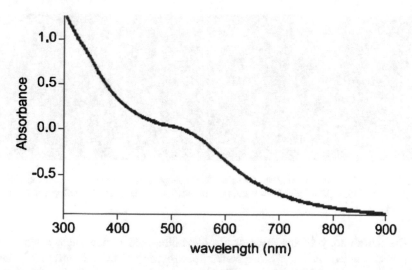

Fig. 7.10 Optical absorption spectrum of a solution of gold nanoparticles, shown in Fig. 7.8 [453]. (Reprinted from Berzina et al. [453], Copyright (2011), with permission from Elsevier)

As it is clear from Fig. 7.9, the average size of gold nanoparticles is about 50 nm in the case of aniline, as a stabilizing agent, 8–10 in the case of lipoic acid, 3–5 nm in the case of dodecane thiol, and 6–8 nm in the case of octane thiol. In addition, in cases of aniline and lipoic acid, we have seen aggregation of nanoparticles, while in cases of dodecane and octane thiols, regular arrays of monodisperse particles have been observed.

Studies of electrical characteristics of samples, prepared from polyaniline with the addition of synthesized particles, revealed characteristics very similar to those of pure polyaniline. The expected suppression of the conductivity at small values of the applied voltages, resulted from the Schottky effect, due to the significant difference of work functions of gold and polyaniline, has not been observed. All voltage-current characteristics had a linear behavior. This negative result can be explained by the presence of functional groups, preventing direct contact of polyaniline with gold. In this respect, nanoparticles, functionalized according to the scheme, shown in Fig. 7.8 seem to be more interesting. Functionalizing groups in this case make a doping effect on polyaniline, which can provide more close contact of this material with gold nanoparticles.

Optical absorption spectrum of the solution of gold nanoparticles, shown in Fig. 7.8, is presented in Fig. 7.10. The shape of the spectrum is typical for gold nanoparticles solution.

SEM images of dried cast solutions of these nanoparticles, obtained at different magnifications, are shown in Fig. 7.11.

As it is clear from Fig. 7.11, the average size of the particles is about 12 nm. However, imaging with higher magnification (Fig. 7.11) revealed that each spot in Fig. 7.11 a contains four particles with characteristic sizes of about 3–4 nm.

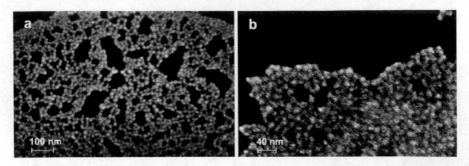

Fig. 7.11 SEM images of dried solutions of gold nanoparticles, shown in Fig. 7.8 at different magnification [453]. (Reprinted from Berzina et al. [453], Copyright (2011), with permission from Elsevier)

The morphology of the composite polyaniline-gold nanoparticle material has been also studied by energy dispersive spectroscopy (EDS). Figure 7.12 shows SEM image (a) and EDS spectrum (b) of this composite material. Particle size in this case was found to be about 25 nm, which can be connected to the particle aggregation during the preparation of the composite material.

As it is clear from Fig. 7.12, the system has a fibrillar structure, similar to the case of vacuum treated material, with the addition of gold nanoparticles, evidenced also by the presence of adequate peaks in Fig. 7.12b [453].

As the solubility of the prepared composite materials in organic solvents was found to be rather low, samples for studying electrical properties in this case were prepared by pressing the material in the form of tablets. Scheme of the experimental sample is shown in Fig. 7.13. The tablets were pressed by the top glass plate to the bottom one, where the system of interdigitated electrodes was formed (distance between electrodes was 50 μ).

Even without doping, the samples revealed high conductivity, which is connected to the use of a special stabilizing agent, acting also as a dopant. In addition, gold nanoparticles themselves can act as doping agents, as is illustrated in Fig. 7.14.

Typical voltage-current characteristics of samples, composed from the described composite materials without any other treatment, are shown in Fig. 7.15. The electrical conductance was found to be rather high: measurements of successive cycles for each sample revealed the decrease of the conductivity not more than 5%.

Figure 7.15 demonstrates the nonlinear properties of the fabricated material. Suppression of the conductivity in the range of low applied voltages is connected to the presence of gold nanoparticles. Despite the fact that the conductivity of gold is much higher than that of polyaniline, they do not form a continuous layer, and there are always gaps between them. Thus, nanoparticles are immersed into a polyaniline matrix, as is shown in Fig. 7.12a. The Schottky barrier appears due to two reasons: close contact of gold and polyaniline and significant difference in work functions of these materials. Therefore, gold nanoparticles input into the total conductance of the material can be reached only when the applied voltage allows to overcome the Schottky barrier.

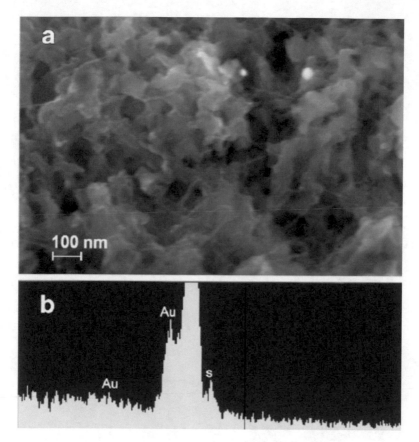

Fig. 7.12 SEM image (**a**) and EDS (**b**) spectrum of polyaniline-gold nanoparticle composite material. (Reprinted from Berzina et al. [453], Copyright (2011), with permission from Elsevier)

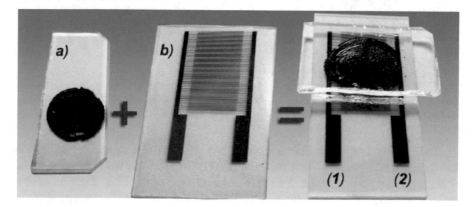

Fig. 7.13 Scheme of the sample, used for studying voltage-current characteristics of the composite material. (Reprinted from Berzina et al. [453], Copyright (2011), with permission from Elsevier)

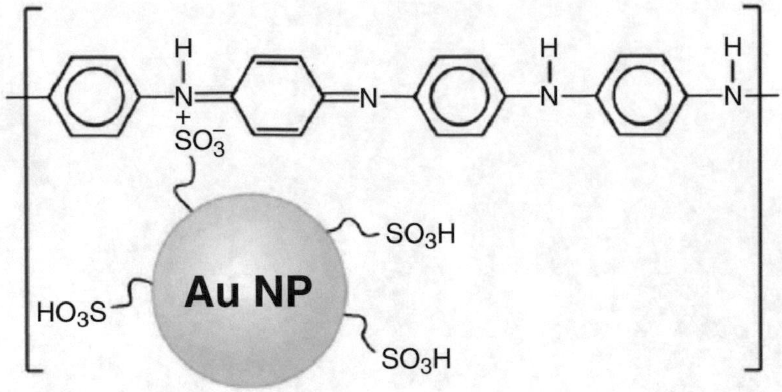

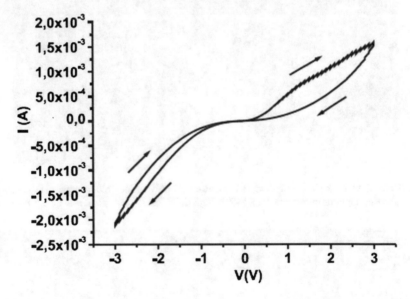

Fig. 7.14 Scheme, illustrating possible doping effect of gold nanoparticles on polyaniline [453]. (Reprinted from Berzina et al. [453], Copyright (2011), with permission from Elsevier)

Fig. 7.15 Cyclic voltage-current characteristics of the composite material without additional doping. Arrows show the direction of the variation of the applied voltage [453]. (Reprinted from Berzina et al. [453], Copyright (2011), with permission from Elsevier)

There is one more interesting feature of the characteristics, shown in Fig. 7.15 – the presence of hysteresis. This behavior can be connected to the accumulation of charges on the particles with their successive drain during the application of the voltage of adequate polarity. Such behavior has been observed in pentacene films with distributed gold particles [458]. It has been shown that holes (the conductivity of pentacene is of p type, as in the case of most organic semiconductors) can be trapped by gold nanoparticles. As the result, hysteresis has been observed when this

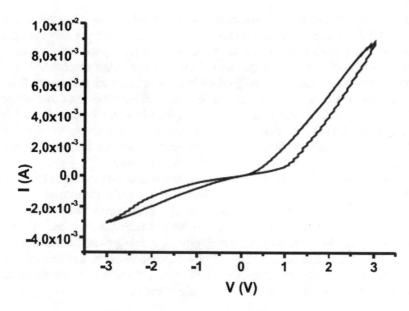

Fig. 7.16 Cyclic voltage-current characteristics of the sample, shown in Fig. 7.13, after additional doping with hydrochloric acid [453]. (Reprinted from Berzina et al. [453], Copyright (2011), with permission from Elsevier)

material was placed between two metal electrodes. Similarly, we can suggest that also in this case, we deal with hole trapping (polyaniline has also p-type conductivity), resulting in the particles charging. Due to the difference in work functions, charging and discharging of particles occur at different values of the applied potential, resulting in hysteresis in voltage-current characteristics.

Typical cyclic voltage-current characteristics of the structure, shown in Fig. 7.13 after doping, is shown in Fig. 7.16.

Comparison of characteristics of undoped (Fig. 7.15) and doped (Fig. 7.16) samples reveals several important points. First, doping results only in one order of magnitude increase of the conductivity. It is to note that in the case of layers, composed from polyaniline only, such doping results in eight orders of magnitude increase of the conductivity. Therefore, the presence of gold nanoparticles themselves and functional groups on their surface provides an already doping effect on polyaniline. Small increase of the sample conductivity can de due to the doping with HCl polyaniline regions that were not in a direct contact with gold nanoparticles. In addition, the presence of gold nanoparticles makes more difficult structural reorganization of the system during doping. Second, hysteresis is maintained after doping, indicating that the Schottky barrier is also maintained. Third, hysteresis after doping is less pronounced. It is connected to the fact that polyaniline itself is more conducting after doping, which facilitates the drain of charges, trapped in the particles, diminishing, therefore, the hysteresis characteristics. Forth, doped samples reveal rectifying behavior. This behavior is not completely clear and can be connected to two factors: variation of the polyaniline layer morphology near gold nanoparticles that can result in the anisotropy of the conductivity and/or entrapment of charges in particles which results in the significant potential difference with adjacent polyaniline layers. In the second case, if the value of potential is high enough, local redox reactions in polyaniline zones can vary their resistance. This last hypothesis is confirmed by the fact that experimentally observed conductance is higher in the positive voltage branch when polyaniline must be in the oxidized conducting state. However, the rectification factor in this case is only about the value of 3, allowing to conclude that only restricted areas of polyaniline, in close contact with gold nanoparticles, are involved in the redox reactions.

These redox reactions can also be an additional reason for the reduced hysteresis: these reactions involve electrons that can be trapped in gold nanoparticles.

The material described in this section is a significant part of more complicated composite material, that will be discussed in the successive subsections. An important property of this material is the possibility of forming Schottky barriers at the boundary with polyaniline, key material of organic memristive devices, due to the significant difference in work function. Thus, these gold nanoparticles are expected to serve as threshold elements in the stochastic network.

7.3.2   Block Copolymer

A block copolymer poly(styrene sulfonic acid)-b-poly-(ethylene oxide)-b-poly(styrene sulfonic acid) (PSS-b-PEO-b- PSS) was prepared following the protocol for a block of PEO obtaining higher molecular weights to ensure greater morphological stability of the 3D matrix assembly.

The preparation of this component is fundamental for the formation of the 3D polymeric network of PEO-PANI and functionalized gold nanoparticles used in organic memristors. A simple route to achieve a stable microphase separation between two components is the preparation of a block copolymer. In this respect, tri-block copolymers PS-b-PEO-b-PS, with a higher molecular weight of the central PEO block, were prepared and then sulfonated. The synthesis consisted of the living radical copolymerization of styrene by atom transfer radical polymerization on the difunctional PEO macroinitiator, prepared by reacting the hydroxy end-groups of PEO and 2-chloro-2-phenylacetylchloride. The resulting tri-block copolymer was sulfonated by reaction of benzene rings with acetyl sulfate in controlled conditions to obtain the PSS-b-PEO-b-PSS. The preparation steps are shown in Fig. 7.17.

These tri-block copolymers are polymeric doping agents for polyaniline with the formation of a microphase-separated material, as is schematically shown in Fig. 7.18.

This method allowed an easy way to modulate the PEO block length (molecular weight) to achieve the desired microphase separated morphology, thus avoiding the macrosegregation of the two polymeric components which are thermodynamically immiscible. It is important to note that this microphase separation yields the

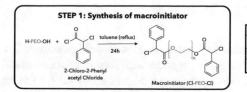

Fig. 7.17 Preparation of a copolymer PS-b-PEO-b-PS and its sulfonation (PSS-b-PEO-b-PSS) [379]. (Reproduced from Erokhin et al. [379] with permission from The Royal Society of Chemistry)

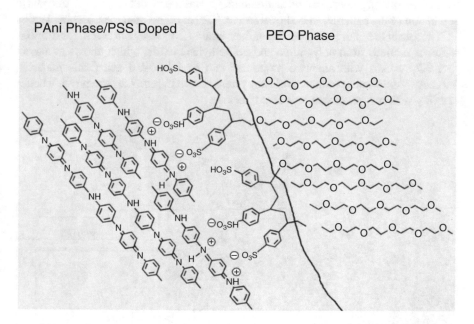

Fig. 7.18 Scheme of the phase separation in the system, based on the copolymer and polyaniline [379]. (Reproduced from Erokhin et al. [379] with permission from The Royal Society of Chemistry)

interfaces between PANI-like and PEO-like areas which can be considered as the microscopic equivalent of the memristive device.

7.3.3   Fabrication of 3D Stochastic Network

Stochastic 3D networks containing polyaniline with gold nanoparticles and block copolymer were assembled on glass supports with four evaporated metal electrodes [379]. After the deposition, part of the material was removed for the realization of cross-like configuration, as is shown in Fig. 7.19.

A ring of adhesive Kapton with a thickness of 36 μ was placed on the sample in such a way that the whole active cross-like zone was within this ring. The well within the ring was filled with electrolyte, containing ions of lithium ($LiClO_4$) and protons (HCl). All redox reactions will occur in this zone, restricted by the ring. In addition, the ring protects metal electrodes from direct contact with electrolytes, which can result in undesirable chemical reactions on their surface.

Three silver wires, playing the role of reference electrodes, were placed on the surfaces of the ring, being in a contact with electrolytes, but not with a composite material under it.

Before making electrical measurements, it has been demonstrated the phase separation of the material, as is shown in Fig. 7.20 at micro- and macroscopic levels.

The observed morphologies are rather similar to those observed for fibrillar systems, realized from polyaniline and polyethylene oxide, which allows to expect that 3D systems with adaptive properties can be fabricated using this material. Moreover, the compact structure of the system shown in Fig. 7.20 can provide better stability with respect to free-standing fibers.

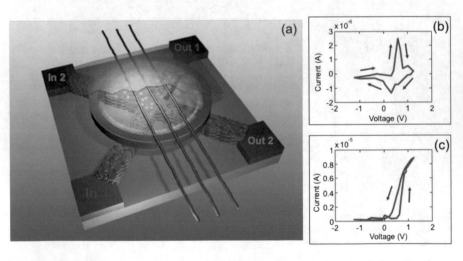

Fig. 7.19 Scheme of the system used for the learning experiments (**a**) and typical cyclic voltage-current characteristics for ionic (**b**) and electronic (**c**) conductivity measured between each input-output pair. Maximum (at about +0.5 V) and minimum (at about +0.1 V) of the ionic current correspond to the oxidation and reduction potentials of PANI, respectively. As a result, the increase or decrease of electronic conductivity is observed. The presence of hysteresis indicates the memory effect in the system [379]. (Reproduced from Erokhin et al. [379] with permission from The Royal Society of Chemistry)

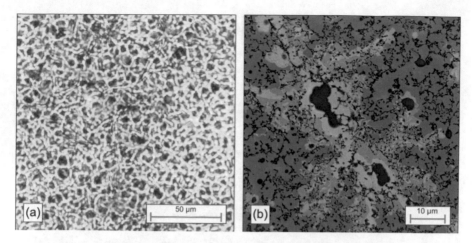

Fig. 7.20 Images of the composite material in the active zone, obtained by optical (**a**) and electron (**b**) microscopes [379]. (Reproduced from Erokhin et al. [379] with permission from The Royal Society of Chemistry)

Similarly to the fibrillar systems, cyclic voltage-current characteristics between different pairs of electrodes have been measured. Typical characteristics for ionic and electronic currents are shown in Fig. 7.19 (a) and (b), respectively.

Comparison of the characteristics, shown in Fig. 7.19, with those of a single organic memristive device allows to conclude that, at least at the level "one input-one output," the system is similar to a discrete deterministic element.

7.3.4   Training of Stochastic 3D Network, Based on Phase Separation of Materials

As the training task, it was decided to induce a high conductivity between one diagonal pair of electrodes (In1-Out1 in Fig. 7.19) and to suppress the conductivity between the other one (In2-Out2 in Fig. 7.19). Two training algorithms have been applied: sequential and simultaneous ones [379].

In the case of sequential algorithm, the procedure was the following: initially, the reinforcing voltage of +0.8 V was applied between one pair of electrodes (In1-Out1 in Fig. 7.19), registering the variation of the conductivity between these electrodes. After the conductivity reached the saturation level, this pair of electrodes was disconnected, and the voltage of −0.2 V was applied between the other pair of electrodes (In2-Out2 in Fig. 7.19). Kinetics of the currents variations for both cases are shown in Fig. 7.21 a and b, respectively.

Similarly to a discrete organic memristive device, kinetics have a different character for positive and negative applied voltages.

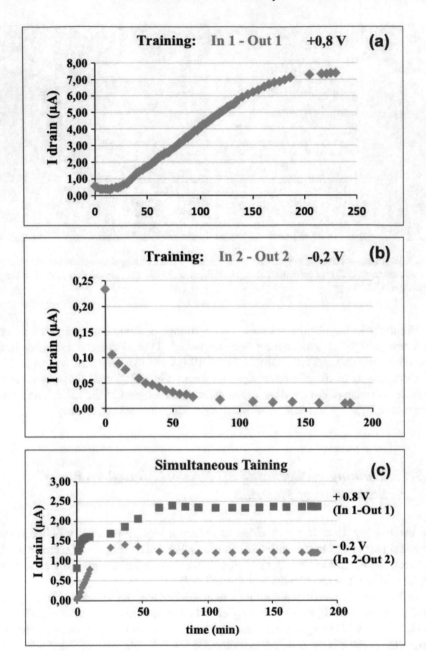

Fig. 7.21 Kinetics of the current variation during the sequential training between In1-Out1 (**a**) and In2-Out2 (**b**) pairs of electrodes, shown in Fig. 7.19; Kinetics of both currents variations during the simultaneous training (**c**) [379]. (Reproduced from Erokhin et al. [379] with permission from The Royal Society of Chemistry)

Table 7.3 Results of 3D stochastic network training (Reproduced from V. Erokhin, T. Berzina, K. Gorshkov, P. Camorani, A. Pucci, L. Ricci, G. Ruggeri, R. Sigala, and A. Schuz, "Stochastic hybrid 3D matrix: learning and adaptation of electrical properties," J. Mater. Chem., 22, 22,881–22,887 (2012) with permission from The Royal Society of Chemistry)

Training			Sequential algorithm I (nA)	Simultaneous algorithm I (nA)
First training	In1-Out1	Immediately	400	200
		After 2 hours	370	250
	In2-Out2	Immediately	7	60
		After 2 hours	5	20
Second training	In1-Out1	Immediately	170	60
		After 2 hours	250	40
	In2-Out2	Immediately	150	300
		After 2 hours	120	370

In order to check the training, after its finishing, a testing voltage of +0.4 V was applied to each pair of electrodes, analyzing the resultant current values. As it was discussed above, the value of this voltage does not vary the conductivity state. Testing results obtained just after training and after 2 hours are presented in Table 7.3.

The data in Table 7.3 demonstrate that the training was successful and the conductivity ratio between different pairs of diagonal electrodes is about 70.

The aim of the second experiment was to retrain the network in such a way that the previously potentiated connection between one pair of electrodes must be depressed, while the initially depressed one must be potentiated. In other words, during the second training, we expected to have a high conductivity between In2-Out2 pair and a low one between In1-Out1 pair. The applied algorithm was similar to that used during the first training.

The results of the second training were rather unexpected: it was impossible to retrain the system. Only small increase of the conductivity between the In2-Out2 pair of electrodes and small decrease of the conductivity between the In1-Out1 pair of electrodes have been observed. These results are also presented in Table 7.3.

In addition, test measurements for the second training were done also one and 2 days after the training. Samples were disconnected when measurements were not performed.

It is interesting to note that in the case of the In1-Out1 pair of electrodes, spontaneous increase of the conductivity has been observed even without any external action. Moreover, after 2 days, the conductivity returned to the value, obtained after the first training (400 nA current value at +0.4 V applied voltage).

The summary of all results, presented above, has demonstrated that the sequential training results in the formation of stable long-term signal pathways. The formed pathways can be only slightly modified during successive training stages. In the absence of external stimuli, the system tends to return to the initial state, reached during the first training.

For explaining the observed behavior, it is necessary to consider the structure of this network that involves zones of conducting polymer, solid electrolyte, insulator, and gold nanoparticles. Therefore, it is possible to consider the system as an array of randomly distributed organic memristive devices with the dispersion of initial electrical properties, resulted from their stochastic nature, immersed into the insulator medium, where gold nanoparticles, capable to trap and release charges, are distributed. As was discussed in the section, dedicated to the cross-talk of devices, the formation of stable signal pathways requires balancing of electrical properties of memristive devices, constituting these pathways. The formation of these pathways involves ion motion between polyaniline and polyethylene oxide zones, which can result in charge accumulation in some zones. Thus, the map of the potential profile in the network, connected also to the local charge distribution, will depend not only on the initial system structure but also on the applied algorithm of its training. In addition, the local variation of maps of the potential distribution can be connected to the charges, trapped on gold nanoparticles [458]. Therefore, long-term sequential training results in not only variation of the polyaniline conductivity in some zones but also in the formation of a rather stable charge distribution maps (consequently, the potential distribution maps) through the whole system. These stable states of the charge and potential distribution act against the significant variations of the system properties during the variation of external stimuli (not very strong and not very long). Moreover, when new external stimuli are switched off, the system tends to return to its initial state.

The next set of experiments was connected to the simultaneous training when the variation of the conductivity of signal pathways between the same pairs of electrodes was done within the same time interval. Both output electrodes were grounded, while voltages of different polarities (similar to sequential training) were applied to the input electrodes. Temporal variations of the current values between these different pairs of electrodes are shown in Fig. 7.21c.

Comparison of characteristics, presented in Fig. 7.21, reveals several differences for the two applied training algorithms. In the case of the sequential training (Fig. 7.21a, b), we can see successive increase of the conductivity between one pair of electrodes (where the positive voltage was applied) and more fast decrease of the conductivity between the second pair of electrodes (where the negative voltage was applied). Instead, in the case of simultaneous training (Fig. 7.21c), we can see the increase of the current value between both pairs of electrodes. This result is connected to the fact that currents in the system in this case are not only between input and output electrodes but also between two input electrodes.

Testing of the efficiency of training was done similarly to the case of sequential training. The results are also shown in Table 7.3.

Testing results have demonstrated that training was successful also in this case. Immediately after the training, the conductivity ratio between two signal pathways was about 4 but reached a value of 12 after 2 hours, indicating the continuation of internal stabilizing processes in the network also in the absence of external stimuli.

Similarly to the case of sequential training, discussed above, the second training procedure was applied to the system (also in a simultaneous manner). This second training tended to invert the conductivity state of signal pathways, stabilized during the first training. Results of these experiments, shown in Table 7.3, have demonstrated that the training in this case was successful.

The difference of results, obtained during sequential and simultaneous training of the system, can be explained in the following way. During each particular moment of sequential training, we can see the formation of a single signal pathway (connection between one pair of electrodes, even if this pathway can have developed structure and contain several linear chains of the signal pathway). In addition, stable maps of the charge and potential distribution are formed in the system during training that can be varied only in the case of the significant variation of amplitudes of externally applied stimuli (in the case of very high amplitudes, it can even destroy completely the network). In the case of simultaneous training, instead, we have two processes going simultaneously in each time moment: reinforcement of the conductivity between one pair of electrodes (creation of occasional signal pathways) and suppression of the conductivity between the other pair of electrodes (elimination of signal pathways). Very likely, it implies that the system is in a dynamic equilibrium: processes responsible for the creation and elimination of signal pathways are always present in the active zone medium. These processes occur simultaneously, forming and destroying connections between different adjacent zones. These processes do not allow to establish stable long-term connections between the pair of electrodes, where we planned to induce the signal pathways, but they also prevent to completely destroy pathways between the other pair of electrodes (spontaneous creation and destruction of signal pathways are always present). In addition, charge and potential distribution maps have also dynamic nature in this case, which also diminish the efficiency of training (conductivity ratio between reinforced and suppressed signal pathways). As a result, the conductivity ratio between pairs of electrodes, obtained during sequential training, is one order of magnitude higher than that obtained during the simultaneous training, but the last one allows effective secondary training of the system.

Qualitative explanation of the results on the training of stochastically organized networks and their connection to the learning of animals and humans can be found in [459].

7.3.5   Evidence of 3D Nature of the Realized Stochastic System

However, an important question about the organization of the network, described in previous subsections, remained: does it really have a 3D organization? In order to

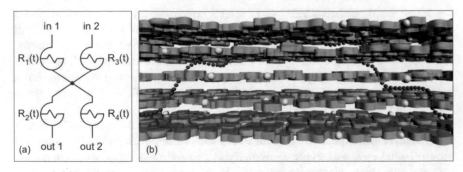

Fig. 7.22 Simplified representation of the possible 2D organization of the active zone (**a**) and the formation of possible 3D signal pathways (red circles) in the hypothetical network, formed due to the self-organization of the used materials (**b**). (Reproduced from Erokhin et al. [379] with permission from The Royal Society of Chemistry)

check it, similar experiments were performed on similar systems, but with the difference that the channel was fabricated from a mono-component film of polyaniline. As in the previous case, measured cyclic voltage-current characteristics, obviously, were similar to those for individual memristive devices. However, training experiments revealed the impossibility of reaching a high conductivity between one pair of diagonal electrodes and a low one between the other pair of electrodes, applying both sequential and simultaneous training algorithms. In the case of sequential training, the state of the conductivity was determined by the last training process: if it was a potentiation phase, high conductivity was observed for both pairs of electrodes; if it was a depression phase, low conductivity was observed for both pairs of electrodes. In the case of simultaneous training, low conductivity was always observed for both pairs of electrodes.

Therefore, these results can be considered as the first demonstration of the fact that the capability of learning and the observed difference of the system features after two applied algorithms can really be connected to the 3D structure of the realized stochastic system, due to the ability to form 3D self-assembling properties of the synthesized copolymer molecules.

In the central active zone of the sample, phase separation of used materials occurs, resulting in the formation of the system that can be considered as a stochastic network of randomly connected organic memristive devices, allowing associative learning. However, the presented data do not allow to state explicitly that the system has a 3D organization. In principle, the potentiation and depression of conductivity in these crossing signal pathways can be reached also in 2D systems. In order to illustrate this statement, let us consider a schematic representation of the active zone, shown in Fig. 7.22a.

Each organic memristive device is characterized by its actual value of the resistance ($R(t)$) in each time moment, dependent on the applied training algorithm, values of external stimuli, learning history, and internal activity of the system. Of

Table 7.4 Test results of the stochastic network learning, demonstrating its 3D organization. (Reproduced from V. Erokhin, T. Berzina, K. Gorshkov, P. Camorani, A. Pucci, L. Ricci, G. Ruggeri, R. Sigala, and A. Schuz, "Stochastic hybrid 3D matrix: learning and adaptation of electrical properties," J. Mater. Chem., 22, 22,881–22,887 (2012) with permission from The Royal Society of Chemistry)

In-Out	Measured current value (A)
In1-Out2	3.1×10^{-6}
In1-Out1	10.2×10^{-9}
In2-Out1	3.6×10^{-6}
In2-Out2	0.28×10^{-6}

course, a real system contains more than four devices, shown in Fig. 7.22a. However, let us consider this scheme for simplicity.

If our task was to increase the conductivity between In1 and Out2 contacts while decreasing it between In2 and Out1 contacts, as is shown in Fig. 7.22a, it will be enough to decrease values of $R_1(t)$ and $R_4(t)$, increasing simultaneously values of $R_2(t)$ and $R_3(t)$. Thus, the results of training, discussed above, can be, in principle, obtained also in the case of the 2D organization of the system. On the other hand, the configuration when we will have a high conductivity between In1-Out2 and In2-Out1 (diagonal pairs of electrodes) and, at the same time, low conductivity between In1-Out1 and In2-Out2 (lateral pairs of electrodes) cannot be realized in the 2D system, as elements involved into the planar array will be used for both diagonal and lateral signal pathways formation. Realization of the described situation requires the possibility of using the third dimension. Therefore, the realization of the situation when the conductivity between diagonal electrodes is much higher than that between lateral ones (or vice versa) would be a direct evidence of the 3D nature of the realized network.

In order to make this verification, the training algorithm has been modified in the following way. During the first training stage, the potentiation of the conductivity (application of +1.0 V) between In1 and Out2 and simultaneous depression of the conductivity (application of −0.6 V) between In1 and Out1 have been done. During the second training state, the reinforcement of the conductivity between In2 and Out1 (application of +1.0 V) and depression of the conductivity (application of −0.6 V) between In2 and Out2 were performed. For testing, the voltage of +0.4 V was applied between different pairs of electrodes, and values of the resultant currents were measured for 6 minutes. Testing results are presented in Table 7.4.

The performed training algorithm has resulted in the potentiation of the conductivity between diagonal pairs of electrodes and its suppression between lateral pairs of electrodes. As has been mentioned above, such connections are not possible in a 2D system. Therefore, these results provide a direct evidence that the realized network, shown in Fig. 7.22b, has a 3D organization.

Summarizing the presented results, it is possible to conclude that mimicking adaptive processes with learning (plasticity) in the nervous system requires the use of elements with properties, similar to those of synapses. For this reason, a composite material, containing stochastically distributed heterojunctions with synapse-mimicking properties and, therefore, including main material components of single organic memristive devices, has been realized and tested. In addition, this composite material contained also gold nanoparticles, serving as threshold elements due to the formation of Schottky barriers, resulted from the significant difference in the work function of polyaniline and gold. These particles (with formed barriers) were considered, therefore, as elements, mimicking in a simplified way the properties of the neuron body (soma). The presence of insulator zones is also very important, as they make divisions of distributed stochastically formed junctions of contacts of polyaniline and solid electrolyte, with properties of organic memristive devices.

It seems interesting to compare the obtained features with some properties of the nervous system and brain. Contrary to computers of the same classes and versions, human and animal brains cannot be considered as a system with precisely defined and identical architectures. Even if there are several similar features, the brain of the animal even of the same type has its own characteristic features. Therefore, even if there are several genetically predetermined schemes of its organization, the brain of each individuum has its characteristic internal connections between elements. Therefore, it can be considered, to a certain degree, as a stochastic distribution of neurons and connections between them. In particular, this statement is correct for the organization of the mammalian brain cortex, the main "instrument," responsible for learning capabilities. Connections between neurons are established through synapses and can be adjusted due to their plasticity: the main reason for the learning capability. Learning is responsible for the functional "structuring" of the brain due to the potentiation and/or depression of some signal pathways.

The behavior of the realized network during the sequential training can be associated with "baby learning" or imprinting. During childhood, strong long-term connections between different areas in the brain, connected to the causality of observed phenomena, are established that can be maintained during the whole life period. Deep learning demands the concentration on one particular association during a significant period of time. At the same time, imprinting implies some depression of connections that are not used during this phase of learning. Sequential training of the described stochastic network has demonstrated practically these features.

Simultaneous training of the described system can be compared with learning during the "everyday" life of an adult person, requiring resolving of problems according to the presence of external stimuli and accumulated experience. It results in the formation of short-term associations that can be varied (if they were not repeated rather frequently), if the external stimuli were changed.

In the brain, the mechanism of "indelible memory" [460] is connected to the elimination of several connections and the potentiation of the other ones. In different parts of the brain, maxima of the connection's formation occur at different ages, followed by the cutting of some branches of axons and synapses. The degree of cutting depends on the richness of the surrounding world: the maximum of this cutting occurs during the sensorial lack in the critical age [460–463]. Therefore, learning during the critical age plays a fundamental role in the formation of characteristic features, determining the individuum, as they make a contribution to the formation of the adult anatomic network. Further learning will vary only connections within the formed anatomic network.

Thus, in the case of sequential training of the network, we realize the formation of stable long-term channels of the signal pathways, which require also stable distribution of trapped charges and potential maps. These stable configurations prevent modifications of connections even in the presence of external stimuli (of course, if they are not very strong and do not repeat rather often). Instead, when we apply the simultaneous algorithm, we do not form stable signal pathways, but only potentiate or depress some of the multiple possible pathways that are in the network, resulting in the formation of short-term associations and memory.

7.4 Modeling of Adaptive Electrical Characteristics of Stochastic 3D Network

The nervous system has capabilities to adaptations, learning, and performing complicated operations (classification, decision-making, forecasting, etc.) with a very high degree of efficiency. Key requirement for these capabilities is the self-organization of the system [464]. As was discussed in the previous section, specially synthesized block copolymers have also the capability to self-organization, allowing the realization of networks with stochastically distributed memristive (synapse-mimicking) junctions, allowing adaptations and learning.

In this section, we will consider the modeling of the dynamics of stochastically organized self-organized systems, based on organic memristive devices, using approaches previously developed and used for the description of similar properties of biological objects. The nature of such systems (both biological and synthetic, described in this chapter) implies the lack of the predetermined structure and a random map of the distribution of connections between nodes (neurons in nature, threshold elements in artificial systems) and memristive devices, mimicking synapses, within these networks that can have a dispersion of structural features and properties.

Considering the stochastic nature of the networks, we will use some parameters (number of nodes, number of connections, map of the connections, etc.), but we will allow the formation of the final configuration (i.e., position of nodes, initial properties of memristive devices, connections, etc.) in a statistic manner. We will focus the attention on understanding the role of connections on the adaptive properties of the network (supervised and unsupervised learning), with a particular aim of reproducing experimentally obtained results and testing the 3D nature of these networks.

The model of the memristive network has a modular organization with different levels, starting from a single memristive device to a complex system with numerous interconnected elements [465].

Four items will be discussed in this section: (1) single memristive device, (2) structure of the network, (3) dynamics of the network, and (4) properties of the

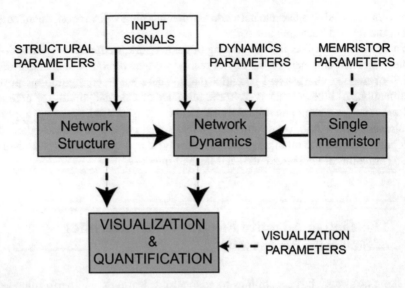

Fig. 7.23 Schematic representation of the modular organization of the model. Module "Network Structure" is responsible for the connections within the network; module "Network Dynamics" simulates the evolution of the network states; module "Visualization and Quantification" calculates global statistics after simulations. Sub-module "Single memristor," describing the behavior of a single device, is used inside the module "Network Dynamics" [465]. (Reproduced with permission from Sigala et al. [465]. Copyright (2013) IOP Publishing)

network. Parameters of single devices will be considered in the first part. The second part will be dedicated to the generation of the 3D organization of nodes and connections between them. The third part will present results on the conductivity of all memristive devices in the network in each time moment, taking into account the results of two first parts and values of the applied external stimuli. Finally, quantitative characteristics of the network activity will be estimated in the fourth part of the section. Each module of the model uses a set of independent parameters. The distribution of modules and their mutual connections within the model are shown in Fig. 7.23.

7.4.1   Single Memristive Device

Due to a large number of devices in the stochastic network, it is difficult to apply directly precise models, considered in Chap. 3, because they require significant calculation time even for describing the functioning of a single memristive element. Therefore, a simplified model of the device function, describing, nevertheless, rather well general characteristics of organic memristive devices, will be used in this section. The main features of this model were considered in Chap. 3. However,

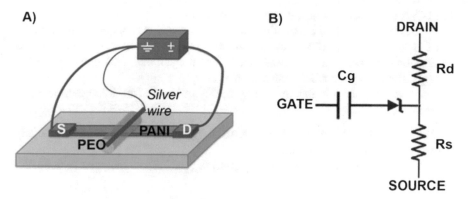

Fig. 7.24 Schematic representation of the organic memristive device (**a**) and equivalent circuit, describing its electrical characteristics (**b**) [465]. (Reproduced with permission from Sigala et al. [465]. Copyright (2013) IOP Publishing)

some particularities will be considered also here for the better understanding of other features of the whole network.

Schematic representation of organic memristive devices and its equivalent circuit are shown in Fig. 7.24 a and b, respectively.

The organic memristive device can be represented as a "star" connection of two variable resistors (two regions of the conducting polyaniline layer in the active zone: one is closer to the source electrode and the second one to the drain electrode) and a capacitor formed at the junction of polyaniline with a solid electrolyte. The presence of the Zener diode takes into account oxidation and reduction potentials, providing additional currents in the capacitor circuit.

In this model, the voltage is distributed between input and output electrodes (drain and source), while the third electrode (gate or reference electrode) is connected to the input electrode. Electrical conductivity of the device is a function of the relative amount of oxidized states of polyaniline k, determined as $k = k_{ox}/k_{max}$ (the ratio of actually oxidized states k_{ox} to the total number of states k_{max}). The contact of polyaniline channel with solid electrolyte is an active zone, where all redox reactions occur, determining the resistance of the device $R(k)$.

Since the relative amount k of oxidized PANI fragments in the active zone of the device determines both the charge on the capacitor and the device resistance $R(k)$, the memristor state can be described at any time by this single parameter.

The differential equation, used to calculate the time evolution of k in our simulation (in the case of oxidation), was defined as follows (Eq. 7.1):

$$dk/dt = (1\text{-}k) \, P(V_j - V_{ox}) \, V_j/R(k) \qquad (7.1)$$

where the temporal derivative of the relative number of oxidized states k (ionic current) is determined as a product of three components:

1. $(1-k)$ is a number of states, that can be oxidized.
2. P – the probability of the oxidation reaction in the active zone, that is the function of both the actual voltage on the active zone with respect to the reference electrode and the oxidation potential V_{ox} of polyaniline. This dependence was approximated by sigmoid, centered on V_{ox}.
3. The last component is determined by the Ohm low for the ionic current in the junction.

The potential of the junction V_j is determined by the applied voltage and the ratio between input and output resistances.

In our case, we used a symmetric equation (eq. 7.2) in the case of reduction processes:

$$\mathrm{d}k/\mathrm{d}t = k\, P\big(V_j - V_{red}\big)\, V_j/R(k) \qquad (7.2)$$

According to the obtained experimental data, the following values for the oxidation and reduction potentials were used: $V_{ox} = +0.4$ V; $V_{red} = +1$ V.

7.4.2 Structure of the Network

To generate the memristor networks, 250 nodes and 1500 connections were distributed within a certain area depending on the simulated conditions.

Connections of nodes in the networks were done according to two different algorithms. To define the connectivity of the networks, we used a so-called "distance-dependent" connectivity, which relates logarithmically the distance between the nodes with the likelihood that they will be connected. Using this rule, networks ended up having a higher number of short connections and a lower number of long ones. For the set of simulations in which we investigated how the adaptive properties of the networks depended on their connectivity pattern, we also used a so-called "fully random" connectivity rule. Figure 7.25 illustrates such distribution of nodes in 3D space for both connectivity rules.

Basing on the experimental results, discussed in Sect. 7.3, it is possible to claim that some parts of the conducting polyaniline areas were not in direct contact with solid electrolyte, which resulted in the appearance of connections with fixed resistance values. In order to take it into account, it was supposed that the probability of resistive (not memristive) connections in the model is about 0.2.

When the network was stochastically generated, two randomly chosen nodes were considered as inputs, and two others randomly chosen nodes were considered as outputs. Two output nodes were connected to the ground through amperemeters, while input nodes were loaded by variable stimuli (voltages). Both for resistors and memristive devices, initial values of the resistance were chosen to be within 50 kΩ – 1 MΩ, characteristic values for the organic memristive devices. Variation of the

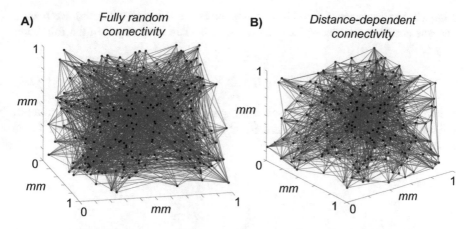

Fig. 7.25 3D representation of memristive network with different distribution of connections. (**a**) Network with fully random connections. (**b**) Network with distance-dependent connections: the probability of the connection between nodes in this model was dependent on the distance between these nodes (as closer, as more probable). The probability decreases logarithmically with the increase of the distance. In both cases, networks modeled to contain 500 nodes and 1500 connections [465]. (Reproduced with permission from Sigala et al. [465]. Copyright (2013) IOP Publishing)

resistance of each memristive device in the network corresponding to the model, discussed in Subsection 7.4.1.

7.4.3 Network Dynamics

To define the dynamics of the networks, the potential distribution at each time step was calculated by solving the circuits, composed of memristive devices and resistors, using the modified nodal analysis (MNA) method. The MNA technique can solve any resistor circuit, independently of the topology, even in the presence of floating current or voltage sources [307, 466]. MNA solves the circuit by calculating the voltage at each circuit node using Kirchhoff's first rule, which states that the algebraic sum of the currents into or out of a node must be equal to the total current injected by current generators, if they are present, in that node.

Calculations were done in the following way.

Equation 7.3 is valid for each ith node:

$$\sum_j I_{ij} = \sum_i \frac{V_i - V_j}{R_{ij}} = \sum_j \frac{1}{R_{ij}} V_i - \sum_j \frac{V_j}{R_{ij}} = I_i^G \qquad (7.3)$$

where summarizing is done for all j nodes, connected to i node through the resistor R_{ij} and I_i^G is a sum of all currents, entering into this node from all current generators, connected to this node ($I_i^G = 0$ if no generator is connected to the node). Each node has one equation, similar to 7.3. These equations can be grouped in the following way (7.4):

$$AV = I, \qquad V = IA^{-1} \tag{7.4}$$

where unknown vector I contains currents, generated by all generators in the circuit, unknown vector V contains potentials at all nodes, and matrix elements A_{ij} is determined by (7.5):

$$A_{ij} = \begin{cases} \sum k \dfrac{1}{R_{ik}} & \text{if } i = j, \\ -\dfrac{1}{R_{ij}} & \text{if } i <> j. \end{cases} \tag{7.5}$$

In other words, in row i of the main diagonal of A, there is a sum of conductivities of all elements, connected to the node i, while a non-diagonal element of A in the position ij is only the conductivity of the element, connecting nodes i and j with the negative sign.

The modified method of the node's analysis can also take into account the presence of generators with a variable voltage. It results in the appearance of a new equation ($V_i - V_j = V_{gen}$) in the system. However, it results in the appearance of a new unknown unit I_k – the current through these nodes. The system can be written as (7.6):

$$GE = K, \qquad E = KG^{-1} \tag{7.6}$$

E is an unknown vector that can contain all potentials in each node, resulting in the appearance of currents in each voltage generator. New vector K contains all currents, entering each node and resulting in the appearance of the potential difference, produced by each generator. New matrix G contains four sub-matrices (7.7):

$$G = \begin{pmatrix} A & B^T \\ B & C \end{pmatrix} \tag{7.7}$$

where A is a matrix, determined by Eq. 7.5. Matrix B contains the same number of columns as the matrix A, and the number of rows corresponds to the number of the voltage generators; all elements of B are equal to zero, except B_{kp}, where positive electrodes of the generators are connected (these elements are equal to 1), and elements B_{kn}, where negative electrodes of generators are connected (these elements are equal to -1).

7.4.4   Modeling of Experimental Results, Obtained on 3D Stochastic Networks

As has been shown in Sect. 7.3, realized stochastic systems have revealed capabilities to adaptations and learning. Moreover, it was shown that some observed properties can be registered only if the systems have 3D organization.

As the first test of the validity of the developed model, we have reproduced the experimentally measured cyclic voltage-current characteristics between different pairs of electrodes within the system. Parameters were chosen corresponding to the experimental conditions, as is shown in Fig. 7.26a. It was supposed that all organic memristive devices are within a restricted active zone. Moreover, as a potential of the reference electrode (ground) was common for all elements, the value of the voltage, used for the calculation of the resistance variation for each element, was averaged between voltages on two nodes, connected by this element. Finally, it was suggested that input and output points are at the boundary of the active zone, as is shown in Fig. 7.26b.

Comparison of the experimentally obtained results (Fig. 7.26c) with the modeling (Fig. 7.26d) revealed a very good coincidence.

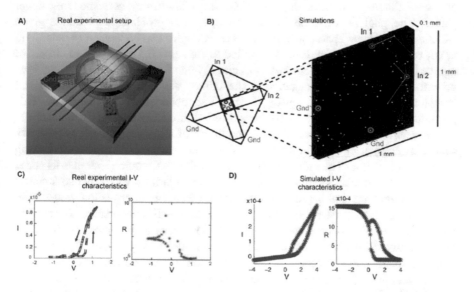

Fig. 7.26 Simulation of the cross-configurated sample, tested in the laboratory. (**a**) Illustration of the experimental setup. (**b**) Schematic view of the experimental configuration and a 3D plot generated during the simulation of the same protocol. The red and blue circles represent inputs and outputs (grounded), respectively. (**c**) I–V characteristics for two opposite electrodes in the real experimental setup. (**d**) I–V characteristics of the simulated network of memristive devices. The left part shows the (drain or electronic) current flowing from one electrode to the opposite one, as a function of the voltage applied to the input nodes. The right part shows the equivalent resistance as a function of the voltage. The hysteresis observed in the system indicates the effect of memory [465]. (Reproduced with permission from Sigala et al. [465]. Copyright (2013) IOP Publishing)

The next step was to simulate general adaptive properties of the 3D network of memristive devices, and not to reproduce specific experimental conditions. During this step, it was suggested, first, that the active area containing all inputs, outputs, and network of memristive devices is within a cub. Second, the reference electrode of each memristive device is connected to one of its terminal electrodes. Third, the position of input and output electrodes was randomly chosen.

Twenty networks were modeled and investigated for each architecture of the network (fully random and distance-dependent). Electrical potentials, applied independently to two pairs of input-output nodes during 100 seconds (epoch), were considered as input signals. The absolute value of the applied potential was chosen randomly within the predefined range, allowing to vary the resistance state of the device for 10–20%, every 20 seconds.

Two types of the input signals were used: (A) signals of the same polarity (with respect to the reference electrode) were applied to both pairs of electrodes; (B) one pair of electrodes was biased with the positive potential difference, while the other one was biased with the negative one. Initial values of the resistances of memristive devices were chosen in a random way, and the whole procedure was repeated 50 times (epochs). Figure 7.27 illustrates the modeling procedure.

In a typical unsupervised learning protocol, a large batch of data belonging to two or more "classes" is used as input of a system without any corresponding target values. The goal of a system in such unsupervised problem is to find a given structure in the input space, such that further examples can be labeled as belonging to any of the given input classes. We applied to the networks two different classes of inputs; in one case, we used competing excitatory and inhibitory signals (A: positive potential in one node and negative in the other), while in the other, we applied complementary reinforcing signals (B: positive potentials in both nodes). This procedure (epoch) was repeated 50 times after reinitializing all resistance values.

For each network, we looked at the evolution of all resistance values (network states), by comparing how similar these values were across the 50 epochs and along the stimulation time. The more adaptive a network is, the more dissimilar its states become as the simulation goes on. To check the similarity between network states, we used a measurement based on correlations, where high correlation values indicating similar network states and low values indicating dissimilar states. For each of the two classes of networks, we compared the similarity across states produced by either the two different types of inputs or by the same class of inputs. The results of these simulations are presented in Fig. 7.28. Correlations on networks with fully random and distance-dependent connectivity are shown in the left and right panels respectively. Each plot shows the average correlation values (S) of 20 different networks as a function of time. Plots on the upper part show correlation values across network states corresponding to different types of input (A versus B). Plots on the bottom part show mean correlation values when the same type of input was used (input A, dashed curve; input B, solid curve). In general, the monotonic and decreasing correlation values as a function of time indicate that the network configurations tend to diverge as longer we stimulate. This behavior is stable regardless of the type of input and network connectivity, which confirm the adaptive properties of

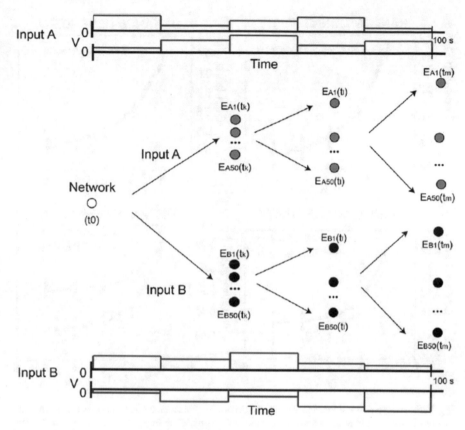

Fig. 7.27 Illustration of the stimulation procedure used to test the adaptive properties of the networks. After having initialized a network (white circle on the left side), we stimulated it during 100 s (epoch duration). As input to the networks, we used two approaches (named A and B), illustrated on the top and bottom parts of the figure. "Input A" implied variations of only positive voltages. "Input B" implied variations of positive and negative voltages. Depending on its initial state and working history, a network was driven to different states, represented by the gray (input A) and black (input B) dots. The procedure, referred to here as an epoch, was repeated 50 times. Network states were vectors with all resistance values and were labeled as E_{Xi} (t_j), x being the type of input, i the epoch number (1:50), and tj the time-step sampled [465]. (Reproduced with permission from Sigala et al. [465]. Copyright (2013) IOP Publishing)

the memristor networks. We observed, however, a difference between fully random and distance-dependent connected networks on how fast their epochs decorrelate, depending on the input used. Networks with a distance-dependent connectivity decorrelated faster when the input was conflicting, i.e. positive in one node and negative in the other (input A), than when both inputs were positive (input B). As observed in Fig. 7.28 (c), this is not the case for the fully randomly connected networks. This indicated that the efficacy of the stimulation depended on both the input pattern and the connectivity of the networks.

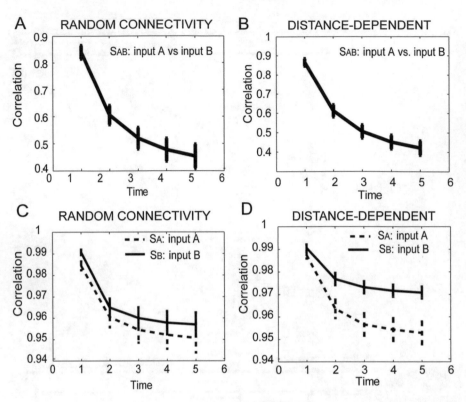

Fig. 7.28 Adaptive behavior in stochastic memristor networks. Evolution of the average similarity as a function of the stimulation time of networks with a fully random (**a, c**) and a distance-dependent connectivity (**b, d**). To calculate the evolution of similarity in a network, all possible pair correlations between network states of different epochs at a given time point were averaged. The decreasing correlation values show how network states deviate more the longer we stimulate. We compared epochs in which different (**a, b**) as well as the same type of inputs (**c, d**) were applied. (**c, d**) Black and dashed curves represent the results according to the type of input used. In contrast to the fully random case, networks with a distance-dependent connectivity seem to deviate faster from the initial state depending on the input used [465]. (Reproduced with permission from Sigala et al. [465]. Copyright (2013) IOP Publishing)

We compared two different kinds of networks that resulted from varying the rules used to define their connectivity: (a) fully randomly connected and (b) distance-dependent connected networks. Networks with a distance-dependent connectivity (dominated by short-range connections) adapted faster to the inputs compared to fully randomly connected networks. It is known that many real networks found in biological, cognitive, and social systems have a distance-dependent connectivity structure. A structure of that kind can be generally described by relating logarithmically the probability of finding a connection between two given nodes and the distance between them, such as by noting that increasing the distance between two nodes reduces the probability that they connect [467]. The fact that many complex

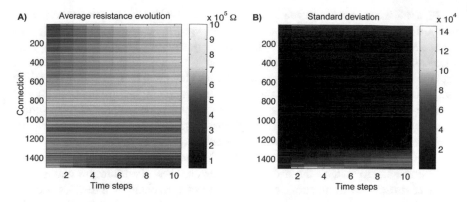

Fig. 7.29 Evolution of the resistance of every connection of a typical network (distance-dependent connectivity) stimulated with the input of type B. The color code represents mean resistance (**a**) and standard deviation (**b**) calculated over 50 epochs. Connections were ranked in the vertical axis according to the difference between the final and the initial resistance. Connections on top became insulating while connections on the bottom became conductive. This figure clearly shows that while the majority of memristive devices became insulating, only a few of them became conductive [465]. (Reproduced with permission from Sigala et al. [465]. Copyright (2013) IOP Publishing)

self-organized systems share similar connectivity structures might indicate an evolutionary advantage reflected in their efficiency and robustness to spread information [468]. In mammalian brains, probably the best example of efficiency and robustness regarding information processing, networks with distance-dependent connectivity also exist: this has been shown for the short-range connectivity within the gray matter of the mammalian cerebral cortex [469, 470] and to a certain degree also for its long-range connectivity via the white matter [471, 472]. The brain and many of its important functions can only be explained by understanding brain connectivity [464]. Although here we present some results that indicate that connectivity influences the adaptive properties of memristor networks, the effect of connectivity in information transmission and learning in these self-organized networks requires further investigation. In particular, it is necessary to study connectivity in relation to other parameters like network size and connection/node ratio.

As indicated by the correlation values, shown in Fig. 7.28d, networks with a distance-dependent connectivity evolved faster when inputs were competitive (input A) than when using two reinforcing inputs (input B). Since the correlation evaluates the resistances of each and every connection in the networks, an interesting question is to look at the proportion of connections that contribute to the differences in the evolution. One possibility is that the differences in the resistances were homogeneously distributed across all the connections. Another possibility is to have some parts of the network more sensitive to features in the input space than others, even with some parts that did not evolve at all. To look at which parts of the networks were sensitive to the inputs applied, we looked at the average change of resistance during stimulation.

Figure 7.29 shows the average behavior in the evolution of each network connection within one epoch, calculated over 50 epochs in which the network was presented with inputs of type B (both voltages positive). The left panel shows the mean resistance values of each link in the network. Each line represents one element, and the lines are ordered according to the average change of resistance. On top of the graph, connections are shown that on average had much higher resistance at the end of the epochs than at the beginning. At the bottom, connections that on average became more conductive during one epoch are presented. The right panel shows the variance for the same evolution, using the same ordering as the left panel.

The black band in the right panel of Fig. 7.29 corresponds to the 20% of resistors in the network for which the resistance never changes, therefore presenting zero variance during the stimulation. A big majority of the memristive devices, located above the black band with zero variance, became on average more insulating. On the other hand, only a relatively small proportion of memristive devices, located below the black band, became more conducting.

When localizing elements of information processing in memristor networks, we distinguished two general properties, as illustrated in Fig. 7.29. The first one is that the memristive devices that become increasingly more insulating are much greater in number than the ones that become more and more conducting. The second one is that the variance in the evolution is much higher in the memristive devices that become more conducting than in the ones that become more insulating. In other words, while each memristive device becomes more insulating more or less in the same manner, every memristive device that on average becomes more conducting does so in very different ways from one epoch to the other. Based on this fact, we can also conclude that information about the input is "stored" in the memristors that become more conductive, which represents the most adaptive part of the networks.

In this section, we have discussed a model to simulate and evaluate the adaptive properties of stochastic memristor networks. In this model, network dynamics could be parametrically simulated with the control at different granularity levels, going from single memristive devices to large-scale configurations. The modular design of this simulation environment allowed us to easily exchange, extend, or substitute any of its elements to perform a wide range of simulations. Here, we first investigated the behavior of the networks recreating real experimental conditions. We have shown that the simulated networks resemble the same hysteretic and rectifying behaviors as the experimental ones. Additionally, we have demonstrated that the simulated networks have indeed a 3D functionality, in contrast to a planar circuit. Finally, in the last group of simulations, we have studied the adaptive properties of three-dimensional (3D) stochastic memristive networks, asking how the connectivity properties of the networks, together with the input, influence their adaptive performance. We have shown that memristor networks stimulated with random noise follow a stable behavior that diverges from their initial state according to the history of stimulation. Interestingly, when varying the connectivity of the networks (from fully random to distance-dependent), we observed differences in the dynamics of the network that depended on the type of input used.

This model was able to capture fundamental qualitative aspects of real self-organized networks of memristive devices. We used a statistical approach to produce a big variability in the structure of the networks, focusing on the general behavior of the matrix. Our model and the simulation environment can contribute to understanding the behavior of complex memristor networks; at the same time, they can help to formulate new hypotheses for future experiments. However, it is important to keep in mind that the model still does not capture important elements of the real experiments, such as the enormous variability among devices, given the particular conditions present in every experiment. Every device is different, and every step of its implementation can be the source of variability.

Conclusions

Direct comparison of features of organic and inorganic devices for possible applications reveals that inorganic memristive devices have obvious advantages for traditional memory arrays: there are well-established technologies for their fabrication, they are stable and can work over a wide temperature range. However, in the case of neuromorphic applications, organic systems have several advantages, described in previous chapters. Comparison of some important properties of organic and inorganic systems is presented in Table A.1.

In my opinion, neuromorphic systems can be roughly represented by the scheme presented in Fig. A.1. The main unit of the system is a processor with memory (or in memory), where the information not only is recorded but also varies connections of processing elements, allowing learning at the hardware level. The system must have several input and output electrodes. This unit must provide parallel information processing, treating simultaneously all the available information coming from input electrodes. It can have both deterministic and/or stochastic architecture. In the first case, the arrays can be fabricated using traditional techniques (lithography, printing, etc.). Such approach can reproduce particular circuits, performing functions of some parts of nervous systems of living beings, when adequate models are available, as it was discussed in Chap. 6, dedicated to the bio-mimicking circuits. The second case seems more neuromorphic and interesting, because it has some similarities with the brain organization [465]. It is to note that the realization of such systems seems to be possible only with organic materials, thanks to their capabilities to self-organization and self-assembling. Systems with stochastic architecture were fabricated using vacuum-induced fiber formation [450], including the utilization of stabilizing support skeleton substrates with developed (porous) surfaces [435], using electrospinning technique [477], and by phase separation of specially synthesized block copolymers [379, 459]. In principle, such systems can be even considered as a combination of multilayer artificial neural networks with a reservoir computer [478, 479]. The last one can provide additional possibilities for the effective processing of temporal sequences of input stimuli.

© Springer Nature Switzerland AG 2022
V. Erokhin, *Fundamentals of Organic Neuromorphic Systems*,
https://doi.org/10.1007/978-3-030-79492-7

Table A.1 Comparison of parameters of inorganic and organic memristive devices

Memristive Devices		
Parameters	Type	
	Inorganic	Organic
Weight	Medium/high	Low
Cost	Medium/high	Low
Flexibility	Low	High
Transparency	Low	High
Compatibility with CMOS technology	High	Low/medium
Endurance (cycles)	10^5 [473] – 10^7 [114]	10^4 [377] – 10^{12} [474]
Biocompatibility	Low	<Medium/high
ON/OFF ratio	10^4 [475] – 10^6 [476]	10^5 [249]

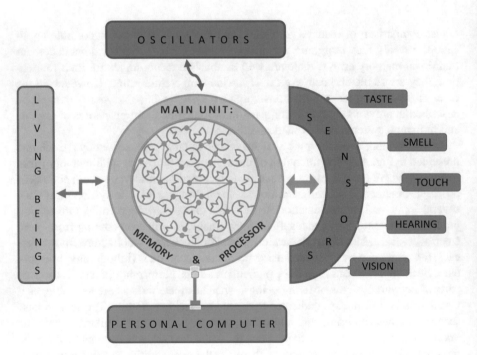

Fig. A.1 Schematic representation of neuromorphic system [480]. (Reprinted by permission from Springer, Erokhin [480], Copyright (2020))

The other essential part of the neuromorphic system is an oscillator. It can provide synchronization of different processing events as well as be an essential part of the "cognitive" activity, when the connections inside the "processor" can be varied even in the case of very low and even absent external stimuli. In the case of organic systems, based on memristive devices, oscillating units can be directly integrated into the central unit, where recording and processing of the information occur. Similarly to living beings, the performance of the oscillating unit (frequency, amplitude) must not be fixed and must depend on the internal activity within the central unit and variation of external stimuli from the seniority systems.

As it was mentioned in the introduction, sensor elements can be based on both inorganic and organic memristive devices. The important feature of these sensors is that the preliminary treatment of the information is done directly on the sensitive element. In particular, it can be performed a direct comparison of the current values of the measured signal with those, obtained in previous moments.

A neuromorphic system must provide detection of signals that animals and humans do. The distribution of research efforts corresponds well to the importance of the information from particular sensory organs, mainly in humans. Most of the works on memristive sensor systems are dedicated to vision detection [173, 481–484]. The second important information, again, in the case of humans, is connected to acoustic signals and voice understanding. Therefore, memristive devices are considered also for the realization of such sensors together with decoding systems [485–489]. It is possible to see rather intensive research activity also in the field of the application of memristive devices for tactile sensors [490–493]. The activity in this field is dedicated also to the necessity of realizing smart actuating systems, such as an artificial hand. Regarding sensing, the worth situation is in the fields of taste and olfactory sensors. There are only few works, mainly describing possibilities, but not really fabricated systems [494, 495]. However, some simplified steps have been done, realizing special sensor systems with memory for detecting elements, molecules, and compounds [178, 183, 184, 493–505].

The other important feature of the neuromorphic system is the possibility of interfacing with living beings. As it was described above in Chap. 6, the interfacing can be done in different ways: readout of signals generated by the nervous system with its successive decoding for the coupling with artificial neural networks [412], modulation of the artificial synaptic strength by neurotransmitters from living beings [413], and direct coupling of live neuronal cells through memristive devices [414]. In the last case, it was done a first step towards the realization of synapse prosthesis. In principle, it is possible to consider even implanted prosthesis. In this case, the issue of biocompatibility is a very critical one. Therefore, special attention was dedicated to the search of materials that are biocompatible, on the one hand, and allow the realization of memristive devices, on the other hand [117, 506, 507]. It is to note that also from the biocompatibility point of view, organic materials seem to be among the best candidates for the fabrication of memristive devices that must be interfaced with biological materials [508–510].

The neuromorphic system can have, of course, an interface with traditional silicon chip-based computers. Such interfacing will eliminate several advantages of neuromorphic systems, such as parallel processing of information and energy consumption. Therefore, it is better to focus efforts on the realization of completely different systems that do not require interfacing with traditional computers. However, currently, it seems impossible to avoid such interfacing. Thus, inorganic materials are widely used for the memristive devices fabrication as their production can be compatible with well-developed CMOS technology, which can simplify the short-term realization of products that can have industrial and commercial impact [511].

It is to note that the realization of complicated information processing systems requires considering also a power consumption factor. The main reason for it is connected to the power dissipation that can result in the decreased stability due to the Joule heating. Of course, it is difficult to make a direct comparison of all inorganic and organic devices because each type of them has its particular characteristics. However, generally speaking, organic electronic devices usually have lower power consumption and, therefore, can be considered as very good candidates also from this point of view.

Finally, there are two other important items: the influence of noise and mutual communication of elements in the network. Regarding the noise influence [340–357, 512], I want to attract the attention of readers to a special issue of "Chaos Solitons and Fractals" (Prof. Bernardo Spagnolo, guest editor) that will be dedicated to noise phenomena in memristive systems. This issue is planned to be published at the end of 2020, and it will provide new systematic information on the noise effects in memristive devices.

Cross-talk of discrete elements in the system plays absolutely different roles in traditional electronics devices and neuromorphic systems. In the case of traditional computers, this role is absolutely negative [513, 514], because the presence of noise will result in the decreased stability and reproducibility of the main performances of the system and generates computational errors. Instead, in the case of neuromorphic systems, mimicking brain properties, this feature is very important [339, 515], because discrete elements are working not independently, but in assembly, when an effect of the damage of individual elements can be eliminated by other elements, performing stable balanced signal pathways. Moreover, cross-talk of elements is also important for the understanding and modeling of cognitive processes, when the brain activity is going on, even if the external stimuli from sensoristic systems are fixed. In this respect, organic systems have an important advantage, as self-assembling techniques allow to build systems with a very high degree of integration and stochastic architecture.

In the end, I want to discuss one other important feature, not considered in this book, that brain has, and computers do not: emotions. Emotions, of course, make a significant influence on the cognitive process [516, 517]. Emotions are directly connected to the presence of neurotransmitters and, as it was suggested, can be roughly represented by the "cub of emotions" [518]. There are several works [519–521], where models, linking the presence of neurotransmitters (emotions), were connected to the redistribution of computational power among memory, the velocity of the information treatment, etc. It is suggested that even learning algorithms can be changed according to the present emotional state. However, the task seems very difficult, and up to now, no hardware realization of computational systems, allowing switching between algorithms of information treatment and learning, has been reported. Nevertheless, I think that this way is very challenging, and hope that works in this direction must be continued.

References

1. N. Rochester, J.H. Holland, L.H. Habit, W.L. Duda, Tests on a cell assembly theory of the action of the brain, using a large digital computer. IRE Trans. Inf. Theory **2**, 80–93 (1956)
2. F. Rosenblatt, The perceptron: A probabilistic model for information storage and organization in the brain. Physiol. Rev. **65**, 386–408 (1958)
3. D.E. Rumelhart, G.E. Hinton, R.J. Williams, Learning internal representation by error propagation, in *Parallel Distributed Processing*, vol. 1, (MIT Press, Cambridge, MA, 1986), pp. 318–362
4. J.J. Hopfield, D.W. Tank, Computing with neural circuits: A model. Science **233**, 625–663 (1986)
5. J.C. Middlebrooks, A.E. Clock, L. Xu, D.M. Green, A panoramic code for sound location by cortical-neurons. Science **264**, 842–844 (1994)
6. H. Cruse, T. Kindermann, M. Schumm, J. Dean, J. Schmitz, Walknet - A biologically inspired network to control six-legged walking. Neural Netw. **11**, 1435–1447 (1998)
7. D.O. Hebb, *Organization of Behavior* (Wiley, New York, 1949), p. 335
8. M.M. Taylor, The problem of stimulus sstructure in the behavioural theory of perceptron. S. Afr. J. Psychol. **3**, 23–45 (1973)
9. W.B. Levy, O. Steward, Temporal contiguity requirements for long-term associative potentiation/depression in the hippocampus. Neuroscience **8**, 791–797 (1983)
10. Y. Dan, M.M. Poo, Hebbian depression of isolated neuromuscular synapses in vitro. Science **256**, 1570–1573 (1992)
11. D. Debanne, B. Gahwiler, S. Thompson, Asynchronous pre- and postsynaptic activity induces associative long-term depression in area CA1 of the rat hippocampus in vitro. Proc. Natl. Acad. Sci. U. S. A. **91**, 1148–1152 (1994)
12. H. Markram, J. Lubke, M. Frotscher, B. Sakmann, Regulation of synaptic efficacy by coincidence of postsynaptic Aps and EPSPs. Science **275**, 213–215 (1997)
13. G.Q. Bi, M.M. Poo, Synaptic modifications in cultured hippocampal neurons: Dependence on spike timing, synaptic strength, and postsynaptic cell type. J. Neurosci. **18**, 10464–10472 (1998)
14. E. Schrödinger, *What is Life? Physical Aspect of the Living Cell* (Cambridge University Press, Cambridge, 1944)
15. R.L. Goldstone, Perceptual learning. Annu. Rev. Psychol. **49**, 585–612 (1998)
16. G.K. Beauchamp, J.A. Mennella, Early flavor learning and its impact on later feeding behavior. J. Pediatr. Gastroenterol. Nutr. **48**, S25–S30 (2009)

© Springer Nature Switzerland AG 2022
V. Erokhin, *Fundamentals of Organic Neuromorphic Systems*,
https://doi.org/10.1007/978-3-030-79492-7

17. E. Di Giorgio, J.L. Loveland, U. Mayer, O. Rosa-Salva, E. Versace, G. Vallortigara, Filial responses as predisposed and learned preferences: Early attachment in chicks and babies. Behav. Brain Res. **325**, 90–104 (2017)
18. L.O. Chua, Memristor – The missing circuit element. IEEE Trans. Circuit Theory **18**, 507–519 (1971)
19. L.O. Chua, S.M. Kang, Memristive devices and systems. Proc. IEEE **64**, 209–223 (1976)
20. F. Corinto, P.P. Civalleri, L.O. Chua, A theoretical approach to memristor devices. IEEE J. Emerg. Selected Topics Circuits Syst. **5**, 123–132 (2015)
21. S. Vongher, X. Shen, The missing memristor has not been found. Sci. Rep. **5**, 11657 (2015)
22. V.A. Demin, V.V. Erokhin, Hidden symmetry shows what a memristor is. Int. J. Unconventional Computing **12**, 433–438 (2016)
23. Y.V. Pershin, M. Di Ventra, A simple test for ideal memristors. J. Phys. D Appl. Phys. **52**, 01LT01 (2019)
24. Y.V. Pershin, M. Di Ventra, Memory effects in complex materials and nanoscale systems. Adv. Phys. **60**, 145–227 (2011)
25. M. Di Ventra, Y.V. Pershin, L.O. Chua, Circuit elements with memory: Memristors, memcapacitors, and meminductors. Proc. IEEE **97**, 1717–1724 (2009)
26. Y.V. Pershin, M. Di Ventra, Memristive circuits simulate memcapacitors and meminductors. Electron. Lett. **46**, 517 (2010)
27. D. Biolek, Z. Biolek, V. Biolkova, SPICE modelling of memcapacitor. Electron. Lett. **46**, 520–521 (2010)
28. D. Biolek, Z. Biolek, V. Biolkova, Behavioral modeling of memcapacitor. Radioengineering **20**, 228–233 (2011)
29. X.Y. Wang, A.L. Fitch, H.H.C. Iu, W.G. Oi, Design of memcapacitor emulator based on a memristor. Phys. Lett. A **376**, 394–399 (2012)
30. C.B. Li, C.D. Li, T.W. Huang, H. Wang, Synaptic memcapacitor bridge synapses. Neurocomputing **122**, 370–374 (2013)
31. B. Liu, B.Y. Liu, X.F. Wang, X.H. Wu, W.N. Zhao, Z.M. Xu, D. Chen, G.Z. Chen, Memristor-integrated voltage-stabilized supercapacitor system. Adv. Mater. **26**, 4999–5004 (2014)
32. M.E. Fouda, A.G. Radwan, Memcapacitor response under step and sinusoidal voltage excitations. Microelectronics J. **45**, 1416–1428 (2014)
33. J.-S. Pei, J.P. Wright, M.D. Todd, S.F. Marsi, F. Gay-Balmaz, Understanding memristors and memcapacitors in engineering mechanics applications. Nonlinear Dynamics **80**, 457–489 (2015)
34. G.Y. Wang, B.Z. Cai, P.P. Jin, T.L. Hu, Memcapacitor model and its application in a chaotic oscillator. Chinese Phys. B **25**, 010503 (2016)
35. J. Mou, K.H. Sun, J.Y. Ruan, S.B. He, A nonlinear circuit with two memcapacitors. Nonlinear Dyn. **86**, 1735–1744 (2016)
36. G. Wang, X. Zang, F. Wang, F. Yuan, H.H.-C. Iu, Memcapacitor model and its application in chaotic oscillator with memristor. Chaos **27**, 013110 (2017)
37. D. Biolek, Z. Biolek, V. Biolkova, PSPICE modeling of meminductor. Analog Integr. Circuits Signal Process. **66**, 129–137 (2011)
38. Y. Liang, D.-S. Yu, H. Chen, A novel meminductor emulator based on analog circuits. Acta Phys. Sinica **62**, 158501 (2013)
39. J. Han, C. Cheng, S. Gao, Y. Wang, C. Chen, F. Pan, Realization of the memcapacitor. ACS Nano **8**, 10043–10047 (2014)
40. F. Yuan, G.-Y. Wang, P.-P. Jin, Study on dynamical characteristics of a memristor model and its meminductor-based oscillator. Acta Phys. Sinica **64**, 210504 (2015)
41. B. Widrow, W.H. Pierce, and J.B. Angell, "Birth, life, and death in microelectronic systems", Technica Report, N. 1552-2/1851-1 (1961).
42. A.O. Bondar, L.A. Dedosha, O.M. Reznik, A.F. Stepanenkov, Simulation of the plasticity of synapses using memistors. Soviet Automatic Control **13**, 47–51 (1968)

43. V. Braitenberg, *Vehicles: Experiments in Synthetic Psychology* (MIT Press, Cambridge, MA, 1984)
44. D. Lambrinos, C. Scheier, "Extended Braitenberg architectures", *Technical Report Al Lab, Computer Science Department, University of Zurich*, N. 95.10 (1995).
45. S. Thakoor, A. Moopenn, T. Daud, A.P. Thakoor, Solid-state thin-film memristor for electronic neural networks. J. Appl. Phys. **67**, 3132–3135 (1990)
46. G.S. Snider, Self-organized computation with unreliable, memristive nanodevices. Nanotechnology **18**, 365202 (2007)
47. V. Erokhin, T. Berzina, M.P. Fontana, Hybrid electronic device based on polyaniline-polyethyleneoxide junction. J. Appl. Phys. **97**, 064501 (2005)
48. D.B. Strukov, G.S. Snider, D.R. Stewart, R.S. Williams, The missing memristor found. Nature **453**, 80–83 (2008)
49. H. Pagnia, N. Sotnik, Bistable switching in electroformed metal-insulator-metal devices. Phys. Status Solidi A **108**, 11–65 (1988)
50. A. Asamitsu, Y. Tomioka, H. Kuwahara, Y. Tokura, Current switching of resistive states in magnetoresistive manganites. Nature **388**, 50–52 (1997)
51. A. Beck, J.G. Bednorz, C. Gerber, C. Rossel, D. Widmer, Reproducible switching effect in thin oxide films for memory applications. Appl. Phys. Lett. **77**, 139 (2000)
52. S.Q. Liu, N.J. Wu, A. Ignatiev, Electric-pulse-induced reversible resistance change effect in magnetoresistive films. Appl. Phys. Lett. **76**, 2749 (2000)
53. R. Waser, M. Aono, Nanoionics-based resistive switching. Nat. Mater. **6**, 833–840 (2007)
54. S. Karthauser, B. Lussem, M. Weides, Resistive switching of rose bengal devices: A molecular effect? J. Appl. Phys. **100**, 094504 (2006)
55. M. Janousch, G.I. Meijer, U. Staub, B. Delley, S.F. Karg, B.P. Andreasson, Role of oxygen vacancies in Cr-doped SrTiO3 for resistance-change memory. Adv. Mater. **19**, 2232–2235 (2007)
56. K. Szot, W. Speier, G. Bihlmayer, R. Waser, Switching the electrical resistance of individual dislocations in single-crystalline $SrTiO_3$. Nat. Mater. **5**, 312–320 (2006)
57. Y.B. Nian, J. Strozier, N.J. Wu, X. Chen, A. Ignatiev, Evidence for an oxygen diffusion models for the electric pulse induced resistance change effect in transition-metal oxides. Phys. Rev. Lett. **98**, 146403 (2007)
58. M. Quintero, P. Levy, A.G. Leyva, M.J. Rozenberg, Mechanism of electric-pulse-induced resistance switching in manganites. Phys. Rev. Lett. **98**, 116601 (2007)
59. X. Chen, N.J. Wu, J. Strozier, A. Ignatiev, Direct resistance profile for an electrical pulse induced resistance change device. Appl. Phys. Lett. **87**, 233506 (2005)
60. M.J. Rozenberg, I.H. Inoue, M.J. Sanchez, Nonvolatile memory with multilevel switching: A basic model. Phys. Rev. Lett. **92**, 178302 (2004)
61. X. Cao, X.M. Li, X.D. Gao, W.D. Yu, X.J. Liu, Y.W. Zhang, L.D. Chen, X.H. Cheng, Forming-free coloccal resistive switching effect in rare-earth-oxide Gd_2O_3 films for memristor applications. J. Appl. Phys. **106**, 073723 (2009)
62. J.J. Yang, F. Miao, M.D. Pickett, D.A.A. Ohlberg, D.R. Stewart, C.N. Lau, R.S. Williams, The mechanism of electroforming of metal oxide memristive switches. Nanotechnology **20**, 215201 (2009)
63. J.J. Yang, M.D. Pickett, X.M. Li, D.A.A. Ohlberg, D.R. Stewart, R.S. Williams, Memristive switching mechanism for metal/oxide/metal nanodevices. Nat. Nanotechnol. **3**, 429–433 (2008)
64. F. Argal, Switching phenomena in titanium oxide thin films. Solid State Electron. **11**, 535–541 (2010)
65. T. Chang, S.H. Jo, K.H. Kim, P. Sheridan, S. Gaba, W. Lu, Synaptic behaviors and modeling of a metal oxide memristive device. Appl. Phys. A **102**, 857–863 (2011)
66. S.E. Savel'ev, A.S. Alexandrov, A.M. Bratkovsky, R.S. Williams, Molecular dynamics simulations of oxide memory resistors (memristors). Nanotechnology **22**, 25401 (2011)

67. M. Cavallini, Z. Hemmatian, A. Riminucci, M. Preziozo, V. Morandi, M. Murgia, Rerenerable resistive switching in silicon oxide based nanojunctions. Adv. Mater. **24**, 1197–1201 (2012)

68. D.B. Strukov, A. Fabien, W.R. Stanley, Thermophoresis/diffusion as a plausible mechanism for unipolar resistive switching in metal-oxide-metal memristors. Appl. Phys. A **107**, 509–518 (2012)

69. A. Younis, D. Adnan, S. Li, Oxygen level: The dominant of resistive switching characteristics in cerium oxide thin films. J. Phys. D Appl. Phys. **45**, 355101 (2012)

70. J. Guo, Y. Zhou, H.J. Yuan, D. Zhao, Y.L. Yin, K. Hai, Y.H. Peng, W.C. Zhou, D.S. Tang, Reconfigurable resistive switching dvices based on individual tungsten trioxide nanowires. AIP Adv. **3**, 042137 (2013)

71. Z.B. Yan, J.-M. Liu, Coexistence of high performance resistance and capacitance memory based on multilayered metal-oxide structures. Sci. Rep. **3**, 2482 (2013)

72. Y. Yang, S.H. Choi, W. Lu, Oxide heterostructure resistive memory. Nano Lett. **13**, 2908–2915 (2013)

73. E. Gale, R. Mayne, A. Adamatzky, B. de Lacy Costello, Drop-coated titanium dioxide memristors. Mater. Chem. Phys. **143**, 524–529 (2014)

74. Y. Aoki, C. Wiemann, V. Feyer, H.-S. Kim, C.M. Schneider, H. Ill-Yoo, M. Martin, Bulk mixed ion electron conduction in amorphous gallium oxide causes memristive behavior. Nat. Commun. **5**, 3473 (2014)

75. S. Kim, S. Choi, W. Lu, Comprehensive physical model of dynamic resistive switching in an oxide memristor. ACS Nano **8**, 2369–2376 (2014)

76. B. Gao, Y.J. Bi, H.Y. Chen, R. Liu, P. Huang, B. Chen, L.F. Liu, J.F. Liu, S.M. Yu, H.S.P. Wong, J.F. Kang, Ultra-low-energy three-dimensional oxide-based electronic synapses for implementation of robust high-accuracy neuromorphic computation systems. ACS Nano **8**, 6998–7004 (2014)

77. V.I. Avilov, O.A. Ageev, A.S. Kolomiitsev, B.G. Konoplev, V.A. Smirnov, O.G. Tsukanova, Formation of a memristor matrix based on titanium oxide and investigation by probe-nanotechnology methods. Semiconductors **48**, 1757–1762 (2014)

78. K. Zhang, Y. Cao, Y. Fang, Q. Li, J. Zhang, C. Duan, S. Yan, Y. Tian, R. Huang, R. Zheng, S. Kang, Y. Chen, G. Liu, L. Mei, Electrical control of memristance and magnetoresistance in oxide magnetic tunnel junctions. Nanoscale **7**, 6334–6339 (2015)

79. M. Prezioso, F. Merrikh-Bayat, B.D. Hoskins, G.C. Adam, K.K. Likharev, D.B. Strukov, Training and operation of an integrated neuromorphic network based on metal-oxide memristors. Nature **521**, 61–64 (2015)

80. E. Gale, D. Pearson, S. Kitson, A. Adamatzky, B. De Lacy Costello, The effect of changing electrode metal on solution-processed flexible titanium dioxide memristors. Mater. Chem. Phys. **162**, 20–30 (2015)

81. A. Younis, D. Chu, S. Li, Evidence of filamentary switching in oxide-based memory devices via weak programming and retention failure analysis. Sci. Rep. **5**, 13599 (2015)

82. A. Regoutz, I. Gupta, A. Serb, A. Khiat, F. Borgatti, T.L. Lee, C. Schlueter, P. Torelli, S. Gobaut, M. Light, D. Carta, S. Pearce, G. Panaccione, T. Prodromakis, Role and optimization of the active oxide layer in TiO2-based RRAM. Adv. Funct. Mater. **26**, 507–513 (2016)

83. M. Prezioso, F.M. Bayat, B. Hoskins, K. Likharev, D. Strukov, Self-adaptive spike-time-dependent plasticity of metal-oxide memristors. Sci. Rep. **6**, 21331 (2016)

84. M. Ungureanu, R. Zazpe, F. Golman, P. Stoliar, R. Llopis, F. Casanova, L.E. Hueso, A light-controlled resistive switching memory. Adv. Mater. **24**, 2496–2500 (2012)

85. M. Prezioso, F.M. Bayat, B. Hoskins, K. Likharev, D. Strukov, Self-adaptive spike-timing-dependent plasticity of metal-oxide memristors. Sci. Rep. **6**, 21331 (2016)

86. G.C. Adam, B.D. Hoskins, M. Prezioso, F. Merrikh-Bayat, B. Chakrabarti, D.B. Strukov, 3-D memristor crossbar for analogand neuromorphic computing applicatios. IEEE trans. Electron Devices **64**, 312–318 (2017)

87. W. Banerjee, Q. Liu, H.B. Lv, S.B. Long, M. Liu, Electronic imitation of behavioral and psychological synaptic activities using TiOx/Al2O3-based memristor devices. Nanoscale **38**, 14442–14450 (2017)

88. G. Baldi, S. Battistoni, G. Attolini, M. Bosi, C. Collini, S. Iannotta, L. Lorenzelli, R. Mosca, J.S. Ponraj, R. Verucchi, V. Erokhin, Logic with memory: AND gates made of organic and inorganic memristive devices. Semiconductor Sci. Technol. **29**, 104009 (2014)

89. X. Cao, X.M. Li, X.D. Gao, W.D. Yu, X.J. Liu, Y.W. Zhang, L.D. Chen, X.H. Chen, Forming-free colossal resistive switching effect in rare-earth-oxide Gd2O3 films for memristor applications. J. Appl. Phys. **106**, 073723 (2009)

90. A. Zimmers, L. Aigouy, M. Mortier, A. Sharoni, S.M. Wang, K.G. West, J.G. Ramirez, I.K. Schuller, Role of thermal heating on the voltage induced insulator-metal transition in VO2. Phys. Rev. Lett. **110**, 056601 (2013)

91. L. Pellegrino, N. Manca, T. Kanki, H. Tanaka, M. Biasotti, E. Bellingeri, A.S. Siri, D. Marre, Multistate devices based on free-standing VO2/TiO2 microstructures driven by Joule self-heating. Adv. Mater. **24**, 2929–2934 (2012)

92. W. Yi, K.K. Tsang, S.K. Lam, X.W. Bai, J.A. Crowell, E.A. Flores, Biological plausibility and stochasticity in scalable VO2 active memristor neurons. Nat. Commun. **9**, 4661 (2018)

93. J. Lappalainen, J. Mizsei, M. Huotari, Neuromorphic thermal-electric circuits based on phase-change VO2 thin-film memristor elements. J. Appl. Phys. **125**, 044501 (2019)

94. J. Sun, I. Maximov, H.Q. Xu, Memristive and memcapacitive characteristics of a Au/Ti-HfO$_2$-InP/InGaAs diode. IEEE Electron Device Lett. **32**, 131–133 (2011)

95. C.H. Kim, J.Y. Byun, J. Kim, M.S. Joo, J.S. Roh, S.K. Park, Dependence of the switching characteristics of resistance random access memory on the type of transition metal oxide: TiO$_2$, ZrO$_2$, and HfO$_2$. J. Electrochem. Soc. **158**, H417–H422 (2011)

96. Y.E. Syu, T.C. Vhang, J.H. Lou, T.M. Tsai, K.C. Chang, M.J. Tsai, Y.L. Wang, M. Liu, S.M. Sze, Atomic-level quantized reaction of HfO$_x$ memristor. Appl. Phys. Lett. **102**, 172903 (2013)

97. A. Wedig, M. Luebben, D.-Y. Cho, M. Moors, K. Skaja, V. Rana, T. Hasegawa, K.A. Adepalli, B. Yildiz, R. Waser, I. Valov, Nanoscale cation motion in TaOx, HfOx and TiOx memristive systems. Nat. Nanotechnol. **11**, 67–74 (2016)

98. Y. Matveyev, R. Kirtaev, A. Fetisova, S. Zakharchenko, D. Negrov, A. Zenkevich, Crossbar nanoscale HfO$_2$-based electronic synapses. Nanoscale Res. Lett. **11**, 147 (2016)

99. H. Jiang, L. Han, P. Lin, Z. Wang, M.H. Jang, Q. Wu, M. Barnell, J.J. Yang, H.L. Xin, Q. Xia, Sub-10 nm Ta channel responsible for superior performance of a HfO$_2$ memristor. Sci. Rep. **6**, 28525 (2016)

100. S. Brivido, J. Frascaroli, S. Spiga, Role of Al doping in the filament distribution in HfO$_2$ resistance switches. Nanotechnology **28**, 395202 (2017)

101. W.F. He, H.J. Sun, Y.X. Zhou, K. Lu, K.H. Xue, X.S. Miao, Coustomized binary and multi-level Hf0$_{2-x}$-based memristors tuned by oxidation conditions. Sci. Rep. **7**, 10070 (2017)

102. Y. Kim, Y.J. Kwon, D.E. Kwon, K.J. Yoon, J.H. Yoon, S. Yoo, H.J. Kim, T.H. Park, J.W. Han, K.M. Kim, C.S. Hwang, Nociceptive memristor. Adv. Mater. **30**, 1704320 (2018)

103. W. Xiong, L. Zhu, C. Ye, F. Yu, Z.Y. Ren, Z.Y. Ge, Bilayered oxide-based cognitive memristor with brain-inspired learning activities. Adv. Electronic Mater. **5**, 1900439 (2019)

104. W.I. Park, J.M. Yoon, M. Park, S.K. Kim, J.W. Jeong, K. Kim, H.Y. Jeong, S. Jeon, K.S. No, J.Y. Lee, Y.S. Jung, Self-assembly-induced formation of high-density silicon oxide memristor nanostructures on graphene and metal electrodes. Nano Lett. **12**, 1235–1240 (2012)

105. A. Younis, D. Chu, X. Lin, J. Yi, F. Dang, S. Li, High-performance nanocomposite based memristor with controlled quantum dots as charge traps, and graphene electrodes. ACS Appl. Mater. Interfaces **5**, 2249–2254 (2013)

106. T. Berzina, K. Gorshkov, V. Erokhin, Y. Nevolin, Y. Chaplygin, Investigation of electrical properties of organic memristors based on thin polyaniline-graphene films. Russ. Microelectron. **42**, 27–32 (2013)

107. Y.C. Yang, J. Lee, S. Lee, C.H. Liu, Z.H. Zhong, W. Lu, Oxide resistive memory with functionalized graphene as built-in selector element. Adv. Mater. **26**, 3693–3699 (2014)

108. S. Porro, C. Ricciardi, Memristive behavior in inkjet printed graphene oxide thin layers. RSC Adv. **5**, 68565–68570 (2015)

109. M. Rogala, P.J. Kowalczyk, P. Dabrowski, I. Wlasny, W. Kozlowski, A. Busiakiewicz, S. Pawlowski, G. Dobinski, M. Smolny, I. Karaduman, L. Lipinska, R. Kozinski, K. Librant, J. Jagiello, K. Grodecki, J.M. Baranowski, K. Szot, Z. Klusek, The role of water in resistive switching in graphene oxide. Appl. Phys. Lett. **106**, 263104 (2015)

110. J. Lee, C. Du, K. Sun, E. Kioupakis, W.D. Lu, Tuning ionic transport in memristive devices by graphene with engineered nanopores. ACS Nano **10**, 3571–3579 (2016)

111. K. Ueda, S. Aichi, H. Asano, Photo-controllable memristive behavior of graphene/diamond heterojunctions. Appl. Phys. Lett. **108**, 222102 (2016)

112. X. Pan, E. Skafidas, Resonant tunneling based graphene quantum dot memristors. Nanoscale **8**, 20074–20079 (2016)

113. H. Tian, W.T. Mi, H.M. Zhao, M.A. Mohammad, Y. Yang, P.W. Chiu, T.L. Ren, A novel artificial synapse with dual modes using bilayer graphene as the bottom electrode. Nanoscale **9**, 9275–9283 (2017)

114. M. Wang, S.H. Cai, C. Pan, C.Y. Wang, X.J. Lian, Y. Zhuo, K. Xu, T.J. Cao, X.Q. Pan, B.G. Wang, S.J. Liang, J.J. Yang, P. Wang, F. Miao, Robust memristors based on layered two-dimensional materials. Nature Electronics **1**, 130–136 (2018)

115. Y. Xin, X.F. Zhao, X.K. Jiang, Q. Yang, J.H. Huang, S.H. Wang, R.R. Zheng, C. Wang, Y.J. Hou, Bistable electrical switching and nonvolatile memory effects by doping different amounts of GO in poly(9,9-dioctylfluorene-2,7-diyl). RSC Adv. **8**, 6878–6886 (2018)

116. Q.Y. Chen, M. Lin, Z.W. Wang, X.L. Zhao, Y.M. Cai, Q. Liu, Y.C. Fang, Y.C. Yang, M. He, R. Huang, Low power parylene-based memristors with a graphene barrier layer for flexible electronics applications. Adv. Funct. Mater. **5**, 1800852 (2019)

117. A.S. Sokolov, M. Ali, R. Riaz, Y. Abbas, M.J. Ko, C. Choi, Silver-adapted diffusive memristor based on organic nitrogen-doped graphene oxide quantum dots (N-GOQDs) for artificial biosynapse applications. Adv. Funct. Mater. **29**, 1807504 (2019)

118. Z.M. Liao, C. Hou, Q. Zhao, D.S. Wang, Y.D. Li, D.P. Yu, Resistive switching and metallic-filament formation in Ag_2S nanowire transistors. Small **5**, 2377–2381 (2009)

119. Y.C. Yang, P. Gao, S. Gaba, T. Chang, X.Q. Pan, W. Lu, Observation of conducting filament growth in nanoscale resistive memories. Nat. Commun. **3**, 732 (2012)

120. D. Li, M.Z. Li, F. Zahid, J. Wang, H. Guo, Oxygen vacancy filament formation in TiO2: A kinetic Monte Carlo study. J. Appl. Phys. **112**, 073512 (2012)

121. C.-H. Huang, J.-S. Huang, C.-C. Lai, H.-W. Huang, S.-J. Lin, Y.-L. Chueh, Manipulated transformation of filamentary and homogeneous resistive switching on ZnO thin film memristor with controllable multistate. ACS Appl. Mater. Interfaces **5**, 6017–6023 (2013)

122. J.Y. Chen, C.L. Hsin, C.W. Huang, C.H. Chiu, Y.T. Huang, S.J. Lin, W.W. Wu, L.J. Chen, Dynamic evolution of conducting nanofilament in resistive switching memories. Nano Lett. **13**, 3671–3677 (2013)

123. L. Zhang, H.Y. Xu, Z.Q. Wang, H. Yu, X.N. Zhao, J.G. Ma, Y.C. Liu, Oxygen-concentration effect on p-type CuAlOx resistive switching behaviors and the nature of conducting filaments. Appl. Phys. Lett. **104**, 93512 (2014)

124. A.J. Lohn, P.R. Mickel, M.J. Marinella, Modeling of filamentary resistive memory by concentric cylinders with variable conductivity. Appl. Phys. Lett. **105**, 83511 (2014)

125. Y.F. Wang, Y.C. Lin, I.T. Wang, T.P. Lin, T.H. Hou, Characterization and modeling of nonfilamentary $Ta/TaO_x/TiO_2/Ti$ analog synaptic device. Sci. Rep. **5**, 10150 (2015)

126. H. Lv, X. Xu, P. Sun, H. Liu, Q. Luo, Q. Liu, W. Banerjee, H. Sun, S. Long, L. Li, M. Liu, Atomic view of filament growth in electrochemical memristive elements. Sci. Rep. **5**, 13311 (2015)

127. J.-Y. Chen, C.-W. Huang, C.-H. Chiu, Y.-T. Huang, W.-W. Wu, Switching kinetic of VCM-based memristor: Evolution and positioning of nanofilament. Adv. Mater. **27**, 5028–5033 (2015)

128. U. Celano, L. Goux, R. Dergaeve, A. Fantini, O. Richard, H. Bender, M. Jurczak, W. Vandervorst, Imaging the three-dimensional conductive channel filamentary-based oxide resistive switching memory. Nano Lett. **15**, 7970–7975 (2015)

129. S. La Barbera, D. Vuillaume, F. Alibart, Filamentary switching: Synaptic plasticity through device volatility. ACS Nano **9**, 941–949 (2015)

130. H. Nakamura, Y. Asai, Competitive effects of oxygen vacancy formation and interfacial oxidation on an ultra-thin HfO_2 based resistive switching memory: Beyond filament and charge hopping models. Phys. Chem. Chem. Phys. **18**, 8820–8826 (2016)

131. Y.-C. Shih, T.H. Wang, J.-S. Huang, C.-C. Lai, Y.-J. Hong, Y.-L. Chueh, Role of oxygen and nitrogen in control of nonlinear resistive behaviors via filamentary and homogeneous switching in an oxynitride thin film memristor. RSC Adv. **6**, 61221–61227 (2016)

132. C. Li, B. Gao, Y. Yao, X.X. Guan, X. Shen, Y.G. Wang, P. Huang, L.F. Liu, X.Y. Liu, J.J. Li, C.Z. Gu, J.F. Kang, R.C. Yu, Direct observation of nanofilament evolution in switching processes in HfO_2-based resistive random access memory by in situ TEM studies. Adv. Mater. **29**, 1602976 (2017)

133. C. Baeumer, R. Valenta, C. Schmitz, A. Locatelli, T.O. Mentes, S.P. Rogers, A. Sala, N. Raab, S. Nemsak, M. Shim, C.M. Schneider, S. Menzel, R. Waser, R. Dittman, Subfilamentary networks cause cycle-to cycle variability in memristive devices. ACS Nano **11**, 6921–6929 (2017)

134. Y. Lu, J.H. Lee, I.-W. Chen, Scalability of voltage-controlled filamentary and nanometallic resistance memory devices. Nanoscale **9**, 12690–12697 (2017)

135. Y. Sun, C. Song, J. Yin, X. Chen, Q. Wan, F. Zeng, F. Pan, Guiding the growth of a conductive filament by nanoindentation to improve resistive switching. ACS Appl. Mater. Interfaces **9**, 34064–34070 (2017)

136. J. Molina-Reyes, L. Hernandez-Martinez, Understanding the resistive switching phenomena of stacked Al/Al2O3/Al thin films from dynamics of conductive filaments. Complexity **2017**, 8263904 (2017)

137. Y. Xia, B. Sun, H. Wang, G. Zhou, X. Kan, Y. Zhang, Y. Zhao, Metal ion formed conductive filaments by redox process induced nonvolatile resistive switching memories in MoS_2 film. Appl. Surf. Sci. **426**, 812–816 (2017)

138. I. Valov, E. Linn, S. Tappertzhofen, S. Schmelzer, J. van den Hurk, F. Lentz, R. Waser, Nanobatteries in redox-based resistive switches require extension of memristor theory. Nat. Commun. **4**, 1771 (2013)

139. Z.Q. Hu, Q. Li, M.Y. Li, Q.W. Wang, Y.D. Zhu, X.L. Liu, X.Z. Zhao, Y. Liu, S.X. Dong, Ferroelectric memristor based on Pt/BiFeO3/Nb-doped SrTiO3 heterostructure. Appl. Phys. Lett. **102**, 102901 (2013)

140. E.Y. Tsymbal, A. Gruverman, Ferroelectric tunnel junctions beyond the barrier. Nat. Mater. **12**, 602–604 (2013)

141. Z.H. Wang, W.S. Zhao, W. Kang, A. Bouchenak-Khelladi, Y. Zhang, Y.G. Zhang, J.O. Klein, D. Ravelosona, C. Chappert, A physics-based compact model of ferroelectric tunnel junction for memory and logic design. J. Phys. D Appl. Phys. **47**, 045001 (2014)

142. S. Boyn, S. Girod, V. Garcia, S. Fusil, S. Xavier, C. Deranlot, H. Yamada, C. Carretero, E. Jacquet, M. Bibes, A. Barthelemy, J. Grollier, High-performance ferroelectric memory based on fully patterned tunnel junctions. Appl. Phys. Lett. **104**, 52909 (2014)

143. A.N. Morosovska, E.A. Eliseev, O.V. Varenyk, Y. Kim, E. Stelcov, A. Tselev, N.V. Morozovsky, S.V. Kalinin, Nonlinear space charge dynamics in mixed ionic-electronic conductors: Resistive switching and ferroelectric-like hysteresis of electromechanical response. J. Appl. Phys. **116**, 066808 (2014)

144. V. Garcia, M. Bibes, Ferroelectric tunnel junctions for information storage and processing. Nat. Commun. **5**, 4289 (2014)

145. L. Kiu, A. Tsurumaki-Fukuchi, H. Yamada, A. Sawa, Ca doping dependence of a resistive switching characteristics in ferroelectric capacitors comprising Ca-doped $BiFeO_3$. J. Appl. Phys. **118**, 204104 (2015)

146. P. Hou, J. Wang, X. Zhong, Y. Wu, A ferroelectric memristor based on the migration of vacancies. RSC Adv. **6**, 54113–54118 (2016)

147. Z.B. Yan, H.M. Yau, Z.W. Li, X.S. Gao, J.Y. Dai, J.-M. Liu, Self-electroforming and high-performance complementary memristor based on ferroelectric tunnel junctions. Appl. Phys. Lett. **109**, 053506 (2016)

148. C.M. Li, Rectification: Light-controlled resistive switching memory of multiferroic BiMnO3 nanowire arrays. Phys. Chem. Chem. Phys. **19**, 10699–10700 (2017)

149. N. Samardzic, B. Bajac, V.V. Srdic, G.M. Stojanovic, Conduction mechanisms in multiferroic multilayer $BaTiO_3/NiFe_2O_4/BaTIO_3$ memristors. J. Electron. Mater. **46**, 5492–5496 (2017)

150. R. Guo, Y.X. Zhou, L.J. Wu, Z.R. Wang, Z.S. Lim, X.B. Yan, W.N. Lin, H. Wang, S.H. Chen, T. Vankatesan, J. Wang, G.M. Chow, A. Gruverman, X.S. Miao, Y.M. Zhu, J.S. Chen, Control of synaptic plasticity learning of ferroelectric tunnel memristor by nanoscale interface engineering. ACS Appl. Mater. Interfaces **10**, 12862–12869 (2018)

151. Z.M. Gao, X.S. Huang, P. Li, L.F. Wang, L. Wei, W.F. Zhang, H.Z. Guo, Reversible resistance switching of 2D electron gas at LaAlO3/SrTiO3 heterointerface. Adv. Mater. Interfaces **5**, 1701565 (2018)

152. B.B. Tian, L. Liu, M.G. Yan, J.L. Wang, Q.B. Zhao, N. Zhong, P.H. Xiang, L. Sun, H. Peng, H. Shen, T. Lin, B. Dkhi, X.J. Meng, J.H. Chu, X.D. Tang, C.G. Duan, A robust artificial synapse based on organic ferroelectric polymer. Adv. Electronic Mater. **5**, 1800600 (2019)

153. F. Xue, X. He, J.R.D. Retamal, A.L. Han, J.W. Zhang, Z.X. Liu, J.K. Huang, W.J. Hu, V. Tung, R.H. He, L.J. Li, X.X. Zhang, Gate-tunable and multidirection-switchable memristive phenomena in a Van der Waals ferroelectric. Adv. Mater. **31**, 1901300 (2019)

154. A. Chanthbouala, V. Garcia, R.O. Cherifi, K. Bouzehouane, S. Fusil, X. Moya, S. Xavier, H. Yamada, C. Deranlot, N.D. Mathur, M. Bibes, A. Barthelemy, J. Grollier, A ferroelectric memristor. Nat. Mater. **11**, 860–864 (2012)

155. M. Ziegler, R. Soni, T. Patelczyk, M. Ignatov, T. Bartsch, P. Meuffels, H. Kohlstedt, An electronic version of Pavlov's dog. Adv. Funct. Mater. **22**, 2744–2749 (2012)

156. V. Erokhin, T. Berzina, P. Camorani, A. Smerieri, D. Vavoulis, J. Feng, M.P. Fontana, Material memristive device circuits with synaptic plasticity: Learning and memory. BioNanoScience **1**, 24–30 (2011)

157. P.R. Benjamin, K. Staras, G. Kemenes, A systems approach to the cellular analysis of associative learning in the pond snail Lymnaea. Learn. Mem. **7**, 124–131 (2000)

158. K. Staras, G. Kemenes, P.R. Benjamin, Pattern-generating role for motoneurons in rhythmically active neuronal network. J. Neurosci. **18**, 3669–3688 (1998)

159. V.A. Straub, P.R. Benjamin, Extrinsic modulation and motor pattern generation in a feeding network: A cellular study. J. Neurosci. **21**, 1767–1778 (2001)

160. M.S. Yeoman, A.W. Pieneman, G.P. Ferguson, A. Ter Maat, P.R. Benjamin, Modulatory role for the sterotonergic cerebral giant cells in the feeding system of the snail, Lymnaea. I. Fine wire recording in the intact animal and pharmacology. J. Neurosci. **72**, 1357–1371 (1994)

161. D.V. Vavoulis, V.A. Straub, I. Kemenes, G. Kemenes, J. Feng, P.R. Benjamin, Dynamic control of a central pattern generator circuit: A computational model of the snail feeding network. Eur. J. Neurosci. **25**, 2805–2818 (2007)

162. E.S. Nikitin, D.V. Vavoulis, I. Kemenes, V. Marra, Z. Pirger, M. Michel, J. Feng, M. O'Shea, P.R. Benjamin, G. Kemenes, Persistent sodium current is a non-synaptic substrate for long-term associative memory. Curr. Biol. **18**, 1221–1226 (2008)

163. D.V. Vavoulis, E.S. Nikitin, I. Kemenes, V. Mara, J. Feng, P.R. Benjamin, G. Kemenes, Balanced plasticity and stability of the electrical properties of a molluscan modulatory interneuron after classic conditioning: A computational study. Front. Behav. Neurosci. **4**, 19 (2010)

164. B.S. Liu, Z.Q. You, X.R. Li, J.S. Kuang, Z. Qin, Comparator and half adder design using complimentary resistive switches crossbar. IEICE Electron. Express **10**, 20130369 (2013)
165. X. Zhu, Y.H. Tang, C.Q. Wu, J.J. Wu, X. Yi, Impact of multiplexed reading scheme on nanocrossbar memristor memory's scalabolity. Chinese Phys. B **23**, 028501 (2014)
166. I. Vourkas, G.C. Sirakoulis, Nano-crossbar memories comprising parallel/serial complementary memristive switches. BioNanoScience **4**, 166–179 (2014)
167. L. Chen, C.D. Li, T.W. Huang, Y.R. Chen, X. Wang, Memristor crossbar-based unsupervised image learning. Neural Comput. Appl. **25**, 393–400 (2014)
168. M. Hu, H. Li, Y. Chen, Q. Wu, G.S. Rose, R.W. Linderman, Memristor crossbar-based neuromorphic computing system: A case study. IEEE Trans. Neural Netw. Learning Syst. **25**, 1864–1878 (2014)
169. M.A. Zidan, H. Omran, A. Sultan, H.A.H. Fahmy, K.N. Salama, Compensated readout for high-density MOS-gated memristor crossbar array. IEEE Trans. Nanotechnol. **14**, 3–6 (2015)
170. M. Wang, X. Lian, Y. Pan, J. Zeng, C. Wang, E. Liu, B. Wang, J.J. Yang, F. Miao, D. Xing, A selector device based on graphene oxide heterostructures for memristor crossbar applications. Appl. Phys. A **120**, 403–407 (2015)
171. C. Yakopcic, R. Hasan, T.M. Taha, Hybrid crossbar architecture for a memristor based cache. Microelectronic J. **46**, 1020–1032 (2015)
172. Q.F. Xia, W. Wu, G.Y. Jung, S. Pi, P. Lin, Y. Chen, X.M. Li, Z.Y. Li, S.Y. Wang, R.S. Williams, Nanoimprint lithography enables memristor crossbar and hybrid circuits. Appl Phys. A **121**, 467–479 (2015)
173. S. Agarwal, T.T. Quach, Q. Parekh, A.H. Hsia, E.P. DeBenedictis, C.D. James, M.J. Marinella, J.B. Aimone, Energy scaling advantages of resistive memory crossbar based computation and its application to sparse coding. Front. Neurosci. **9**, 484 (2016)
174. B.J. Choi, J. Zhang, K. Norris, G. Gibson, K.M. Kim, W. Jackson, M.-X. Zhang, Z. Li, J.J. Yang, R.S. Williams, Alternative architectures toward reliable memristive crossbar memories. Adv. Mater. **28**, 356–362 (2016)
175. M.A. Zidan, H. Omran, R. Naous, A. Sultan, H.A.H. Fahmy, W.D. Lu, K.N. Salama, Single-readout high-density memristor crossbar. Sci. Rep. **6**, 18863 (2016)
176. W. Xu, Y. Lee, S.-Y. Min, C. Park, T.-W. Lee, Simple, inexpensive, and rapid approach to fabricate cross-shaped memristors using an inorganic-nanowire-digital-alignment technique and a one-step reduction process. Adv. Mater. **28**, 527–532 (2016)
177. Y. Li, Y.-Z. Zhou, L. Xu, K. Lu, Z.-R. Wang, N. Duan, L. Jiang, L. Cheng, T.-C. Chang, K.-C. Chang, H.-J. Sun, K.-H. Xue, X.-S. Miao, Realization of functional complete stateful Boolean logic memristive crossbar. ACS Appl. Mater. Interfaces **8**, 34559–34567 (2016)
178. B. Chakrabarti, M.A. Lastras-Montano, G. Adam, M. Preziozo, B. Hoskins, K.-T. Cheng, D.B. Strukov, A multi-add engine with monolithically integrated 3D memristor crossbar/CMOS hybrid circuit. Sci. Rep. **7**, 42429 (2017)
179. C. Li, L.L. Han, H. Jiang, M.H. Jang, P. Lin, Q. Wu, M. Barnell, J.J. Yang, H.L.L. Xin, Q.F. Xia, Three-dimensional crossbar arrays of self-rectifying $Si/SiO_2/Si$ memristors. Nat. Commun. **8**, 15666 (2017)
180. V.A. Demin, V. Erokhin, A.V. Emelyanov, S. Battistoni, G. Baldi, S. Iannotta, P.K. Kashkarov, M.V. Kovalchuk, Hardware elementary perceptron based on polyaniline memristive device. Org. Electronics **25**, 16–20 (2015)
181. O. Kayehei, S.J. Lee, K.R. Cho, S. Al-Sarawi, D. Abbot, A pulse-frequency modulation sensor using memristive-based inhibitory interconnections. J. Nanosci. Nanotechnol. **13**, 3505–3510 (2013)
182. P. Puppo, M. Di Ventra, G. De Micheli, S. Carrara, Memristive sensors for pH measure in dry conditions. Surf. Sci. **624**, 76–79 (2014)
183. F. Puppo, A. Dave, M.A. Douccy, D. Sacchetto, C. Baj-Rossi, Y. Leblebici, G. De Micheli, S. Carrara, Memristive biosensors under varying humidity conditions. IEEE Trans. Nanobioscience **13**, 19–30 (2014)

184. I. Tzouvadaki, F. Puppo, M.A. Doucey, G. De Micheli, S. Carrara, Computational study on the electrical behavior of silicon nanowire memristive biosensors. IEEE Sensors J. **15**, 6208–6217 (2015)
185. I. Tzouvadaki, C. Parrozzani, A. Gallotta, G. De Michele, S. Carrara, Memristive biosensors for PSA-IgM detection. BioNanoScience **5**, 189–195 (2015)
186. I. Tzouvadaki, N. Madaboosi, I. Taurino, V. Chu, J.P. Conde, G. De Micheli, S. Carrara, Study on the bio-functionalization of memristive nanowires for optimum memristive biosensors. J. Mater. Chem. B **4**, 2153–2162 (2016)
187. I. Tzouvadaki, P. Jolly, X. Lu, S. Ingebrandt, G. de Micheli, S. Carrara, Label-free ultrasensitive memristive aptasensor. Nano Lett. **16**, 4472–4476 (2016)
188. B. Ibarlucea, T.F. Akbar, K. Kim, T. Rim, C.K. Baek, A. Ascoli, R. Tetzlaff, L. Baraban, G. Cuniberti, Ultrasensitive detection of Ebola matrix protein in a memristor mode. Nano Res. **11**, 1057–1068 (2018)
189. A. Adeyemo, J. Mathew, A. Jabir, C. Di Natale, E. Martinelli, M. Ottavi, Efficient sensing approaches for high-density memristor sensor array. J. Comput. Electron. **17**, 1285–1296 (2018)
190. K.D. Cantley, A. Subramaniam, H.J. Stiegler, R.A. Chapman, E.M. Vogel, Hebbian learning in spiking neural networks with nanocrystalline silicon TFTs and memristive synapses. IEEE Trans. Nanotechnol. **10**, 1066–1073 (2011)
191. J.-S. Huang, W.-C. Yen, S.-M. Lin, C.-Y. Lee, J. Wu, Z.M. Wang, T.-S. Chin, Y.-L. Chueh, Amorphous zinc-doped silicon oxide (SZO) resistive switching memory: Manipulated bias control from selector to memristor. J. Mater. Chem. C **2**, 4401–4405 (2014)
192. A.N. Mikhailov, A.I. Belov, D.V. Guseinov, D.S. Korolev, I.N. Antonov, D.V. Efimovykh, S.V. Tikhov, A.P. Kasatkin, O.N. Gorshkov, D.I. Tetelbaum, A.I. Bobrov, N.V. Malekhonova, D.A. Pavlov, E.G. Gryaznov, A.P. Yatmanov, Bipolar resistive switching and charge transport in silicon oxide memristor. Mater. Sci. Engineer. B **194**, 48–54 (2015)
193. L. Martinez, D. Becerra, V. Agarwal, Dual layer ZnO configuration over nanostructured porous silicon substrate for enhanced memristive switching. Superlattices Microstruct. **100**, 89–96 (2016)
194. V. Erokhin, M.P. Fontana, Thin film electrochemical memristive systems for bio-inspired computation. J. Comput. Theor. Nanosci. **8**, 313–330 (2010)
195. S. Kim, H.Y. Jeong, S.K. Kim, S.Y. Choi, K.J. Lee, Flexible memristive memory array on plastic substrates. Nano Lett. **11**, 5438–5442 (2011)
196. S.M. Yoon, S. Yang, S.W. Jung, C.W. Byun, M.K. Ryu, W.S. Cheong, B. Kim, H. Oh, S.H. Park, C.S. Hwang, S.Y. Kang, H.J. Ryu, B.G. Yu, Polymeric ferroelectric oxide semiconductor-based fully transparent memristor cell. Appl. Phys. A **102**, 983–990 (2011)
197. M.K. Hota, M.K. Bera, B. Kundu, S.C. Kundu, C.K. Maiti, A natural silk fibroin protein-based transparent bio-memristor. Adv. Funct. Mater. **22**, 4493–4499 (2012)
198. M.N. Awais, K.H. Choi, Resistive switching and current conduction mechanism in full organic resistive switch with the sandwiched structure of poly(3,4-ethylenedioxythiophene): poly(styrenesulfonate)/poly(4-vinylphenol)/poly(3,4-ethylenedioxythiophene):   poly (styrenesulfonate). Electron. Mater. Lett. **10**, 601–606 (2014)
199. Y. Wang, X. Yan, R. Dong, Organic memristive devices based on silver nanoparticles and DNA. Org. Electron. **15**, 3476–3481 (2014)
200. S. Qin, R. Dong, X. Yan, Q. Du, A reproducible write-(read)n-erese and multilevel bio-memristor based on DNA molecule. Org. Electronics **22**, 147–153 (2015)
201. B. Sun, L. Wei, H. Li, X. Jia, J. Wu, P. Chen, The DNA strand assisted conductive filament mechanism for improved resistive switching. J. Mater. Chem. B **3**, 12149–12155 (2015)
202. Y.-C. Chen, H.-C. Yu, C.-Y. Huang, W.-L. Chung, S.-L. Wu, Y.-K. Su, Nonvolatile bio-memristor fabricated with egg albumen film. Sci. Rep. **5**, 10022 (2015)
203. F. Zeng, S.Z. Li, J. Yang, F. Pan, D. Guo, Learning processes modulated by the interface effects in a Ti/conducting polymer/Ti resistive switching cell. RSC Adv. **4**, 14822–14828 (2014)

204. N.R. Hosseini, J.-S. Lee, Resistive switching memory based on bioinspired natural solid polymer electrolytes. ACS Nano **9**, 419–426 (2015)

205. N. Raeis-Hosseini, J.-S. Lee, Controlling the resistive switching behavior in starch-based flexible biomemristors. ACS Appl. Mater. Interfaces **8**, 7326–7332 (2016)

206. Y. Cai, J. Tan, Y.F. Liu, M. Lin, R. Huang, A flexible organic resistance memory device for wearable biomedical applications. Nanotechnology **27**, 275206 (2016)

207. S. Song, J. Jang, Y. Ji, S. Park, T.-W. Kim, Y. Song, M.-H. Yoon, H.C. Ko, G.-Y. Jung, T. Lee, Twistable nonvolatile organic resistive memory devices. Org. Electronics **14**, 2087–2092 (2013)

208. D. Son, S. Qiao, R. Ghaffari, Multifunctional wearable devices for diagnosis and therapy of movement disorders. Nat. Nanotechnol. **9**, 397–404 (2014)

209. R. Wang, Y. Liu, B. Bai, N. Guo, J. Guo, X. Wang, M. Liu, G. Zhang, B. Zhang, C. Xue, Wide-frequency-bandwidth whisker-inspired MEMS vector hydrophone encapsulated with parylene. J. Phys. D Appl. Phys. **49**, 07LT02 (2016)

210. B.J. Kim, C.A. Gutierrez, E. Meng, Parylene-based electrochemical-MEMS force sensor for studies of intracortical probe insertion mechanisms. J. Microelectromech. Syst **24**, 1534–1544 (2015)

211. A.A. Minnekhanov, A.V. Emelyanov, D.A. Lapkin, K.E. Nikiruy, B.S. Shvetsov, A.A. Nesmelov, V.V. Rylkov, V.A. Demin, V.V. Erokhin, Parylene based memristive devices with multilevel resistive switching for neuromorphic applications. Sci. Rep. **9**, 10800 (2019)

212. S. Saighi, C.M. Mayr, T. Serrano-Gotarredona, H. Schmidt, G. Lecerf, J. Tomas, J. Grollier, S. Boyn, A.F. Vincent, D. Querlioz, S. La Barbera, F. Alibard, D. Vuillaume, O. Bichler, C. Gamrat, B. Linares-Barranco, Plasticity in memristive devices for spiking neural networks. Front. Neurosci. **9**, 51 (2015)

213. I.P. Pavlov, Experimental psychology and psychopathology in animals, in *Lectures on Conditioned Reflexes*, vol. 1, (International Publishers, New York, 1928), pp. 47–60

214. P. Dayan, S. Kakade, P.R. Montague, Learning and selective attention. Nat. Neurosci. **3**, 1218–1223 (2000)

215. Z. Wang, M. Rao, J.-W. Han, J. Zhang, P. Lin, Y. Li, C. Li, W. Song, S. Asapu, R. Midya, Y. Zhuo, H. Jiang, J.H. Yoon, N.K. Upadhyay, S. Joshi, M. Hu, J.P. Strachan, M. Barnell, Q. Wu, H. Wu, Q. Qiu, R.S. Williams, Q. Xia, J.J. Yang, Capacitive neural network with neuro-transistors. Nat. Commun. **9**, 3208 (2018)

216. E.T. Kang, K.G. Neoh, K.L. Tan, Polyaniline: A polymer with many interesting intrinsic redox states. Prog. Polym. Sci. **23**, 277–324 (1998)

217. E.W. Paul, A.J. Ricco, M.S. Wrighton, Resistance of polyaniline film as a function of electrochemical potential and the fabrication of polyaniline-based microelectronic devices. J. Phys. Chem. **89**, 1441–1447 (1985)

218. G.B. Appetecchi, F. Alessandrini, M. Carewska, T. Caruso, P.P. Prosini, S. Scaccia, S. Passerini, Investigation on lithium-polymer electrolyte batteries. J. Power Sources **97-98**, 790–794 (2001)

219. V. Erokhin, G. Raviele, J. Glatz-Reichenbach, R. Narizzano, S. Stagni, C. Nicolini, High-value organic capacitor. Mater. Sci. Eng. C **22**, 381–385 (2002)

220. V.I. Troitsky, T.S. Berzina, M.P. Fontana, Langmuir-Blodgett assemblies with patterned conductive polyaniline layers. Mater. Sci. Eng. C **22**, 239–244 (2002)

221. G.G. Roberts, P.S. Vincett, W.A. Barlow, Technological applications of Langmuir-Blodgett films. Phys. Technol. **12**, 69–75 (1981)

222. R.H. Tredgold, The physics of Langmuir-Blodgett films. Rep. Prog. Phys. **50**, 1609–1656 (1987)

223. H. Kuhn, Present status and future prospects of Langmuir-Blodgett film research. Thin Solid Films **178**, 1–16 (1989)

224. V. Erokhin, R. Kayushina, Y. Lvov, N. Zakharova, A. Kononenko, P. Knox, A. Rubin, Preparation of Langmuir films of photosynthetic reaction centers from purple bacteria. Dokl. Akad. Nauk SSSR **299**, 1262–1267 (1988)

225. Y. Lvov, V. Erokhin, S. Zaitsev, Langmuir-Blodgett protein films. Biol. Mem. **7**, 917–937 (1990)
226. V. Erokhin, P. Facci, C. Nicolini, Two-dimensional order and protein thermal stability: High temperature preservation of structure and function. Biosens. Bioelectron. **10**, 25–34 (1995)
227. V. Erokhin, P. Facci, A. Kononenko, G. Radicchi, C. Nicolini, On the role of molecular close packing on the protein thermal stability. Thin Solid Films **284-285**, 805–808 (1996)
228. V. Erokhin, Langmuir-Blodgett multilayers of proteins, in *Protein Architecture: Interfacing Molecular Assemblies and Immobilization Biotechnology*, ed. by Y. Lvov, H. Möhwald, (Marcel Dekker, Inc., New York, 2000), pp. 99–124
229. V. Erokhin, Langmuir-Blodgett films of biological molecules, in *Handbook of Thin Film Materials*, ed. by H. S. Nalwa, vol. 1. Deposition and Processing of Thin Films, (Academic Press, San Diego, 2002), pp. 523–558
230. T. Berzina, V. Erokhin, M.P. Fontana, Spectroscopic investigation of an electrochemically controlled conducting polymer – solid electrolyte junction. J. Appl. Phys. **101**, 024501 (2007)
231. R.P. McCall, J.M. Ginder, J.M. Leng, H.J. Ye, S.K. Manohar, J.G. Masters, G.E. Asturias, A.G. MacDiamid, A.J. Epstein, Spectroscopy and defect states in polyaniline. Phys. Rev. B **41**, 5202 (1990)
232. L. Abell, S.J. Pomfret, P.N. Adams, A.C. Middleton, A.P. Monkman, Studies of stretched predoped polyaniline films. Synth. Met. **84**, 803–804 (1997)
233. F. Pincella, P. Camorani, V. Erokhin, Electrical properties of organic memristive system. Appl. Phys. A **104**, 1039–1046 (2011)
234. N.S. Sariciftci, H. Kuzmany, H. Neugebauer, A. Neckel, Structural and electronic transitions in polyaniline: A Fourier transform infrared spectroscopic study. J. Chem. Phys. **92**, 4530 (1990)
235. S. Nie, S.R. Emory, Probing single molecules and single nanoparticles by surface-enhanced Raman scattering. Science **275**, 1102–1106 (1997)
236. E.J. Blackie, E.C. Le Ru, M. Meyer, P.G. Etchegoin, Surface enhanced Raman scattering enhancement factors: A comprehensive study. J. Phys. Chem. C **111**, 13794–13803 (2007)
237. E.J. Blackie, E.C. Le Ru, P.G. Etchegoin, Single-molecule surface-enhanced Raman spectroscopy of nonresonant molecules. J. Am. Chem. Soc. **131**, 14466–14472 (2009)
238. W.B. Yun, J.M. Bloch, X-ray near total external fluorescence method: Experiment and analysis. J. Appl. Physiol. **68**, 1421 (1990)
239. S.I. Zheludeva, M.V. Kovalchuk, S. Lagomarsino, N.N. Novikova, I.N. Bashelkhanov, V. Erokhin, L.A. Feigin, Observation of evanescent and standing X-ray waves in region of total external reflection from molecular Langmuir-Blodgett films. JEPT Lett. **52**, 170–175 (1990)
240. T. Berzina, S. Erokhina, P. Camorani, O. Konovalov, V. Erokhin, M.P. Fontana, Electrochemical control of the conductivity in an organic memristor: A time-resolved X-ray fluorescence study of ionic drift as a function of the applied voltage. ACS Appl. Mater. Interfaces **1**, 2115–2118 (2009)
241. J. Feng, *Computational Neuroscience. A Comprehansive approach* (Chapman and Hal, CRC, Boca Raton, London, New York, Washington, 2004), p. 640
242. A. Smerieri, T. Berzina, V. Erokhin, M.P. Fontana, Polymeric electrochemical element for adaptive networks: Pulse mode. J. Appl. Phys. **104**, 114513 (2008)
243. E.R. Holland, S.J. Pomfret, P.N. Adams, A.P. Monkman, Conductivity studies of polyaniline doped with CSA. J. Phys. Condens. Matter **8**, 2991–3002 (1996)
244. P.N. Adams, P. Devasagayam, S.J. Pomfret, L. Abell, A.P. Monkman, A new acid-processing route to polyaniline films which exhibit metallic conductivity and electrical transport strongly dependent upon intrachain molecular dynamics. J. Phys. Condens. Matter **10**, 8293–8303 (1998)
245. V. Erokhin, T. Berzina, P. Camorani, M.P. Fontana, On the stability of polymeric electrochemical elements for adaptive networks. Coll. Surf. A **321**, 218–221 (2008)

246. S.V. Ayrapetiants, T.S. Berzina, S.A. Shikin, V.I. Troitsky, Conducting Langmuir-Blodgett films of binary mixtures of donor and acceptor molecules. Thin Solid Films **210-211**, 261–264 (1992)

247. V. Erokhin, T. Berzina, M.P. Fontana, Polymeric elements for adaptive networks. Cryst. Rep. **52**, 159–166 (2007)

248. V. Erokhin, A. Schuz, M.P. Fontana, Organic memristor and bio-inspired information processing. Int. J. Unconventional Computing **6**, 15–32 (2010)

249. T. Berzina, A. Smerieri, M. Bernabo, A. Pucci, G. Ruggeri, V. Erokhin, M.P. Fontana, Optimization of an organic memristor as an adaptive memory element. J. Appl. Phys. **105**, 124515 (2009)

250. Y. Haba, E. Segal, M. Narkis, G.I. Titelman, A. Siegmann, Polymerization of aniline in the presence of DBSA in an aqueous dispersion. Synth. Met. **106**, 59–66 (1999)

251. W.A. Gazotti, M.-A. De Paoli, High yield preparation of a soluble polyaniline derivative. Synth. Met. **80**, 263–269 (1996)

252. T. Berzina, A. Smerieri, G. Ruggieri, M. Bernabo, V. Erokhin, M.P. Fontana, Role of the solid electrolyte composition on the performance of the polymeric memristor. Mater. Sci. Eng. C **30**, 407–411 (2010)

253. G. Decher, Fuzzy nanoassemblies: Toward layered polymeric microcomposites. Science **277**, 1232–1237 (1997)

254. J.H. Cheung, W.B. Stokton, M.F. Rubner, Molecular-level processing of conjugated polymers. 3. Layer-by-layer manipulation of polyaniline via electrostatic interactions. Macromolecules **30**, 2712 (1997)

255. S. Erokhina, V. Sorokin, V. Erokhin, Polyaniline-based organic memristive device fabricated by Layer-by-Layer deposition technique. Electron. Mater. Lett. **11**, 801–805 (2015)

256. D. Braun, P. Fromherz, Fluorescence interferometry of neuronal cell adhesion on microstructured silicon. Phys. Rev. Lett. **81**, 5241 (1998)

257. A.A. Prinz, P. Fromherz, Electrical synapses by guided growth of cultured neurons from snail Lymneaea stagnalis. Biol. Cybern. **82**, L1–L5 (2000)

258. B. Straub, E. Meyer, P. Fromherz, Recombinant maxi-K channels on transistor, a prototype of iono-electronic interfacing. Nat. Biotechnol. **19**, 121–124 (2001)

259. M. Merz, P. Fromherz, Polyester microstructures for topographical control of outgrowth and synapse formation of snail neurons. Adv. Mater. **14**, 141–144 (2002)

260. P. Fromherz, Electrical interfacing of nerve cells and semiconductor chips. Chem. Phys. Chem. **3**, 276–284 (2002)

261. I. Gupta, A. Serb, A. Khiat, R. Zeiter, S. Vassanelli, T. Prodromakis, Real-time encoding and compression of neural spikes by metal-oxide memristors. Nat. Commun. **7**, 12805 (2016)

262. V. Erokin, T. Berzina, P. Camorani, M.P. Fontana, Non-equilibrium electrical behavior of polymeric electrochemical junctions. J. Phys. Condens. Matter **19**, 205111 (2007)

263. N. Zaikin, A.M. Zhabotinsky, Concentration wave propagation in two-dimensional liquid-phase self-oscillating system. Nature **225**, 535–537 (1970)

264. P. Glansdor, I. Prigogine, *Thermodynamic Theory of Structure, Stability and Function* (Wiley, New York, 1971), p. 230

265. S. Komaba, T. Itabashi, T. Kimura, H. Groult, N. Kumagai, Opposite influence of K+ versus Na+ ions as electrolyte additives on graphite electrode performance. J. Power Sources **146**, 166–170 (2005)

266. A. Smerieri, V. Erokhin, M.P. Fontana, Origin of current oscillations in a polymeric electrochemically controlled element. J. Appl. Phys. **103**, 094517 (2008)

267. V.A. Demin, V.V. Erokhin, P.K. Kashkarov, M.V. Kovalchuk, Electrochemical model of the polyaniline based organic memristive device. J. Appl. Phys. **116**, 064507 (2014)

268. B. J. O'M, A. K. N. Reddy, M, Gamboa-Aldeco (eds.), *Modern Electrochemistry. Fundamentals of Electrodics*, vol 2A (Kluwer Academic Publishers, New York, 2002), p. 1534

269. V. Allodi, V. Erokhin, M.P. Fontana, Effect of temperature on the electrical properties of an organic memristive device. J. Appl. Phys. **108**, 074510 (2010)

270. T.F. Fuller, M. Doyle, J. Newman, Relaxation phenomena in lithium-ion-insertion cells. J. Electrochem. Soc. **141**, 982 (1994)

271. D.A. Lapkin, A.N. Korovin, S.N. Malakhov, A.V. Emelyanov, V.A. Demin, V.V. Erokhin, Optical monitoring of the resistive states of a polyaniline-based memristive device. Adv. Electron. Mater **6**, 2000511 (2020)

272. A. Dimonte, F. Fermi, T. Berzina, V. Erokhin, Spectral imaging method for studying Physarum polycephalum growth on polyaniline surface. Mater. Sci. Eng. C **53**, 11–14 (2015)

273. S. Battistoni, A. Dimonte, V. Erokhin, Spectroscopic characterization of organic memristive devices. Org. Electronics **38**, 79–83 (2016)

274. V. Erokhin, G.D. Howard, A. Adamatzky, Organic memristive devices for logic elements with memory. Int. J. Bifurcation Chaos **22**, 1250283 (2012)

275. Y. Levy, J. Bruck, Y. Cassuto, E.G. Friedman, A. Kolodny, E. Yaakobi, S. Kvatinsky, Logic operations in memory using memristive Akers array. Microelectronics J. **45**, 1429–1437 (2017)

276. A. Wu, S. Wen, Z. Zeng, Synchronization control of a class of memristor-based recurrent neural networks. Inform. Sci. **183**, 106–116 (2012)

277. G. Indiveri, B. Linares-Barranco, R. Legenstein, T. Prodramakis, Integration of nanoscale memristor synapses in neuromorphic computing architectures. Nanotechnology **24**, 384010 (2013)

278. L.Q. Zhu, C.J. Wan, L.Q. Guo, Y. Shi, Q. Wan, Artificial synapse network on inorganic proton conductor for neuromorphic systems. Nat. Commun. **5**, 3158 (2014)

279. J.W. Jang, S. Park, G.W. Burr, H. Hwang, Y.H. Jeong, Optimization of conductance change in Pr1-xCaxMnO3-based synaptic devices for neuromorphic systems. IEEE Electron Device Lett. **36**, 457–459 (2015)

280. Y. van de Burgt, E. Lebberman, E.J. Fuller, S.T. Keene, G.C. Faria, S. Agarwal, M.J. Marinella, A.A. Talin, A. Salleo, A non-volatile organic electrochemical device as a low-voltage artificial synapse for neuromorphic computing. Nat. Mater. **16**, 414 (2017)

281. P.D. Wasserman, *Neural Computing: Theory and Practice* (Van Nostrand Reinhold, New York, 1989)

282. F. Rosenblatt, *Principles of Neurodynamics: Perceptrons and the Theory of Brain Mechanism* (Spartan Books, Washington DC, 1961)

283. V.A. Demin, V. Erokhin, A.V. Emelyanov, S. Battistoni, G. Baldi, S. Iannotta, P.K. Kashkarov, M.V. Kovalchuk, Hardware elementary perceptron based on polyaniline memristive devices. Org. Electronics **25**, 16–20 (2015)

284. A.V. Emelyanov, D.A. Lapkin, V.A. Demin, V.V. Erokhin, S. Battistoni, G. Baldi, A. Dimonte, A.N. Korovin, S. Iannotta, P.K. Kashkarov, M.V. Kovalchuk, First step towards the realization of a double layer perceptron, based on organic memristive decices. AIP Adv. **6**, 111301 (2016)

285. S. Battistoni, V. Erokhin, S. Iannotta, Organic memristive devices for perceptron applications. J. Phys. D Appl. Phys. **51**, 284002 (2018)

286. F.M. Bayat, M. Prezioso, B. Chakrabarti, H. Nili, I. Kataeva, D. Strukov, Implementation of multilayer perceptron network with highly uniform passive memristive crossbar circuits. Nat. Commun. **9**, 2331 (2018)

287. H. Jeong, L. Shi, Memristor devices for neural networks. J. Phys. D Appl. Phys. **52**, 023003 (2018)

288. S.P. Wen, S.X. Xiao, Y. Yang, Z. Yan, Z.G. Zeng, T.W. Huang, Adjusting learning rate of memristor-based multilayer neural networks via fuzzy method. IEEE Trans. Comput.-Aided Des. Integr. Circ. Syst. **38**, 1084–1094 (2019)

289. O. Krestinskaya, K.N. Salama, A.P. James, Learning in memristive neural network architectures using analog backpropagation circuits. IEEE Trans. Circ. Syst. I **66**, 719–732 (2019)

290. Z.R. Wang, C. Li, W.H. Song, M.Y. Rao, D. Belkin, Y.N. Li, P. Yan, H. Jiang, P. Lin, M. Hu, J.P. Strachan, N. Ge, M. Barnell, Q. Wu, A.G. Bartos, Q.R. Qiu, R.S. Williams, Q.F. Xia,

J.J. Yang, Reinforcement learning with analogue memristor arrays. Nat. Electron. **2**, 115–124 (2019)

291. G.D. Zhou, Z.J. Ren, L.D. Wang, B. Sun, S.K. Duan, Q.L. Song, Artificial and wearable albumen protein memristor arrays with integrated memory logic gate functionality. Mater. Horizons **6**, 1877–1882 (2019)

292. O. Krestinskaya, A.P. James, L.O. Chua, Neuromemristive circuits for edge computing: A review. IEEE Trans. Neural Netw. Learning Syst. **31**, 4–23 (2020)

293. J. Lin, J.-S. Yuan, A scalable and reconfigurable in-memory architecture for ternary deep spiking neural network with ReRAM based neurons. Neurocomputing **375**, 102–112 (2020)

294. P. Yao, H. Wu, B. Gao, J.S. Tang, Q.T. Zhang, W.Q. Zhang, J.J. Yang, H. Qian, Fully hardware-implemented memristor convolutional neural network. Nature **577**, 641 (2020)

295. F. Silva, M. Sanz, J. Seixas, E. Solano, Y. Omar, Perceptrons from memristors. Neural Netw. **122**, 273–278 (2020)

296. M. Lukosevicius, H. Jaeger, Reservoir computing approaches to recurrent neural network training. Comput. Sci. Rev. **3**, 127–149 (2009)

297. J. Moon, W. Ma, J.H. Shin, F.X. Cai, C. Du, S. Lee, W.D. Lu, Temporal data classification and forecasting using a memristor-based reservoir computing system. Nature Electronics **2**, 480–487 (2019)

298. P.J. Werbos, *The Roots of Backpropagation. From Ordered Derivatives to Neural Networks and Political Forecasting* (John Wiley and Sons, Inc, New York, 1994)

299. N.V. Prudnikov, D.A. Lapkin, A.V. Emelyanov, A.A. Minnekhanov, Y.N. Malakhova, S.N. Chvalun, V.A. Demin, V.V. Erokhin, Associative STDP-like learning of neuromorphic circuits based on polyaniline memristive microdevices. J. Phys. D Appl. Phys. **53**, 414001 (2020)

300. A. Smerieri, T. Berzina, V. Erokhin, M.P. Fontana, A functional polymeric material based on hybrid electrochemically controlled junctions. Mater. Sci. Eng. C **28**, 18–22 (2008)

301. S.G. Hu, Y. Liu, Z. Liu, T.P. Chen, Q. Yu, L.J. Deng, Y. Yin, S. Hosaka, Synaptic long-term potentiation realized in Pavlov's dog model based on a NiOx-based memristor. J. Appl. Phys. **116**, 214502 (2014)

302. L.D. Wang, H.F. Li, S.K. Duan, T.W. Huang, H.M. Wang, Pavlov associative memory in a memristive neural network and its circuit implementation. Neurocomputing **171**, 23–29 (2016)

303. C.X. Wu, T.W. Kim, T.L. Guo, F.S. Li, D.U. Lee, J.J. Yang, Mimicking classical conditioning based on a single flexible memristor. Adv. Mater. **29**, 1602890 (2017)

304. Z. Wang, X. Wang, A novel memristor-based circuit implementation of full-function Pavlov associative memory accorded with biological feature. IEEE Trans. Circuits and Systems I **65**, 2210–2220 (2020)

305. A.A. Minnekhanov, B.S. Shvetsov, M.M. Martyshov, K.E. Nikiruy, E.V. Kukueva, M.Y. Presnyakov, P.A. Forsh, V.V. Rylkov, V.V. Erokhin, V.A. Demin, A.V. Emelyanov, On the resistive switching mechanism of parylene-based memristive devices. Org. Electronics **74**, 89–95 (2019)

306. V. Erokhin, T. Berzina, A. Smerieri, P. Camorani, S. Erokhina, M.P. Fontana, Bio-inspired adaptive networks based on organic memristors. Nano Commun. Netw. **1**, 108–117 (2010)

307. A. Ruehli, P. Brennan, The modified nodal approach to network analysis. IEEE Trans. Circ. Syst. **22**, 504–509 (1975)

308. J.L. Meador, A. Wu, C. Cole, N. Nintinze, P. Chintrakulchai, Programmable impulse neural circuits. IEEE Trans. Neural Netw. **2**, 101–109 (1991)

309. W.C. Fang, B.J. Sheu, O.T.C. Chen, J. Choi, A VLSI neural processor for image data-compression using self-organization networks. IEEE Trans. Neural Netw. **2**, 506–518 (1992)

310. H. Kosaka, T. Shibata, H. Ishii, T. Ohmi, An excellent weight-updating-linearity EEPROM synapse memory cell for self-learning neuron-MOS neural networks. IEEE Trans. Electron Devices **42**, 135–143 (1995)

311. A.J. Montalvo, R.S. Gyurcsik, J.J. Paulos, Toward a general-purpose analog VLSI neural network with on-chip learning. IEEE Trans. Neural Netw. **8**, 413–432 (1997)

312. C. Diorio, D. Hsu, M. Figueroa, Adaptive CMOS. From biological inspiration to systems-on-a-chip. Proc. IEEE **90**, 345–357 (2002)
313. E. Chicca, D. Badoni, V. Dante, M. D'Andreagiovanni, G. Salina, L. Carota, S. Fusi, P. Del Giudice, A VLSI recurrent network of integrate-and-fire neurons connected by plastic synapses with long-term memory. IEEE Trans. Neural Netw. **14**, 1297–1307 (2002)
314. R.J. Vogelstein, U. Mallik, J.T. Vogelstein, G. Cauwenberghs, Dynamically reconfigurable silicon array of spiling neurons with conductance-based synapses. IEEE Trans. Neural Netw. **18**, 253–265 (2007)
315. J.H.B. Wijekoon, P. Dudek, Compact silicon circuit with spiking and bursting behavior. Neural Netw. **21**, 524–534 (2007)
316. K.K. Likharev, CrossNets: Neuromorphic hybrid CMOS/nanoelectronics networks. Sci. Adv. Mater. **3**, 322–331 (2011)
317. J.M. Cruz-Albrecht, M.C.W. Yung, N. Srinivasa, Energy-efficient neuron, synapse and STDP integrated circuits. IEEE Trans. Biomed. Circuits Systems **6**, 246–256 (2012)
318. S. Brink, S. Nease, P. Hasler, S. Ramakrishnan, R. Wunderlich, A. Basu, B. Degnan, A learning-enabled neuron array IC based upon transistor channel models of biological phenomena. IEEE Trans. Biomed. Circuits Systems **7**, 71–81 (2013)
319. N. Qiao, H. Mostafa, F. Corradi, M. Osswald, F. Stefanini, D. Sumislawska, G. Indiveri, A reconfigurable on-line learning spiking neuromorphic processor comprising 256 neurons and 128K synapses. Front. Neurosci. **9**, 141 (2015)
320. C. Mayr, J. Partzsch, M. Noack, S. Hanzsche, S. Scholze, S. Hoppner, G. Ellguth, R.A. Schuffny, A biological-realtime neuromorphic system in 28 nm CMOS using low-leakage switched capacitor circuits. IEEE Trans. Biomed. Circ. Syst. **10**, 243–254 (2016)
321. I. Kemenes, V.A. Straub, E.S. Nikitin, K. Staras, M. O'Shea, G. Kemenes, P.R. Benjamin, Role of delayed nonsynaptic neuronal plasticity in long-term associative memory. Curr. Biol. **16**, 1269–1279 (2006)
322. W. Zhang, D.J. Linden, The other side of the engram: Experience-driven changes in neuronal intrinsic excitability. Nat. Rev. Neurosci. **4**, 885–900 (2003)
323. P.R. Benjamin, G. Kemenes, I. Kemenes, Non-synaptic neuronal mechanisms of learning and memory in gastropod mollusks. Front. Biosci. **13**, 4051–4057 (2008)
324. C.H. Bailey, M. Giustetto, Y.Y. Huang, R.D. Hawkins, E.R. Kandel, Is heterosynaptic modulation essential for stabilizing Hebbian plasticity and memory? Nat. Rev. Neurosci. **1**, 11–20 (2000)
325. W. Robinett, M. Pickett, J. Borghetti, Q.F. Xia, G.S. Snider, G. Medeiros-Ribeiro, R.S. Williams, A memristor-based nonvolatile latch circuit. Nanotechnology **21**, 235203 (2010)
326. C. Moreno, C. Munuera, S. Valencia, F. Kronast, X. Obradors, C. Ocal, Reversible resistive switching and multilevel recording in La0.7Sr0.3Mn0.3 thin films for low cost nonvolatile memories. Nano Lett. **10**, 3828–3835 (2010)
327. M.J. Lee, C.B. Lee, D. Lee, S.R. Lee, M. Chang, J.H. Hur, Y.B. Kim, C.J. Kim, D.H. Seo, S. Seo, U.I. Chung, I.K. Yoo, A. Kim, A fast, high-endurance and scalable non-volatile memory device made from asymmetric Ta2O5-x/TaO2-x bilayer structure. Nat. Mater. **10**, 625–630 (2010)
328. Y.Y. Fang, R.K. Dumas, T.N.A. Nguyen, S.M. Mohseni, S. Chung, C.W. Miller, J. Akerman, A nonvolatile spintronic memory element with a continuum of resistance states. Adv. Funct. Mater. **23**, 1919–1922 (2013)
329. T. Braz, Q. Ferreira, A.L. Mendonca, A.M. Ferraria, A.M.B. do Rego, J. Morgado, Morphology of ferroelectric/conjugated polymer phase-separated blends used in nonvolatile resistive memories. J. Phys. Chem. C **119**, 1391–1399 (2015)
330. Y. Sun, L. Li, D. Wen, X. Bai, G. Li, Bistable electrical switching and nonvolatile memory effect in carbon nanotube-poly(3,4-ethylenedioxythiophene): poly(styrenesulfonate) composite film. Phys. Chem. Chem. Phys. **17**, 17150–17158 (2015)

331. A. Ascoli, R. Tetzlaff, L.O. Chua, J.P. Strachan, R.S. Williams, Hystory erase effect in a non-volatile memristor. IEEE Trans. Circuits Systems I **63**, 389–400 (2016)
332. S. Ali, J. Bae, C.H. Lee, S. Shin, N.P. Kobayashi, Ultra-low power non-volatile resistive crossbar memory based on pull up resistors. Org. Electronics **41**, 73–78 (2017)
333. D. Liu, Q. Lin, Z. Zang, M. Wang, P. Wangyang, X. Tanag, M. Zhou, W. Hu, Flexible all-inorganic perovskite CsPbBr3 nonvolatile memory device. ACS Appl. Mater. Interfaces **9**, 6171–6176 (2017)
334. M. Yang, N. Qin, L.Z. Ren, Y.J. Wang, K.G. Yang, F.M. Yu, W.Q. Zhou, M. Meng, S.X. Wu, D.H. Bao, S.W. Li, Realizing a family of transition-metal-oxide memristors based on volatile resistive switching at a rectifying metal/oxide interface. J. Phys. D. **47**, 045108 (2014)
335. R. Berdan, C. Lim, A. Khiat, C. Papavassiliou, T. Prodromakis, Amemristor SPICE model accounting for volatile characteristics of practical ReRAM. IEEE Electron Device Lett. **35**, 135–137 (2014)
336. J. Van der Hurk, E. Linn, H.H. Zhang, R. Waser, I. Valov, Volatile resistance states in electrochemical metallization cells enabling non-destructive readout of complementary resistive switches. Nanotechnology **25**, 425202 (2014)
337. Y.V. Pershin, S.N. Shevchenko, Computing with volatile memristors: An application of non-pinched hysteresis. Nanotechnology **28**, 075204 (2017)
338. T. Berzina, K. Gorshkov, V. Erokhin, Chains of organic memristive devices: Cross-talk of elements. AIP Conf. Proc. **1479**, 1888–1891 (2012)
339. A. Dimonte, T. Bersina, M. Pavesi, V. Erokhin, Hysteresis loop and cross-talk of organic memristive devices. Microelectronics J. **45**, 1396–1400 (2014)
340. D. Valenti, A. Fiasconaro, B. Spagnolo, Stochastic resonance and noise delayed extinction in a model of two competing species. Phys. A **331**, 477–486 (2004)
341. A. Fiasconaro, B. Spagnolo, Stability measures in metastable states with Gaussian colored noise. Phys. Rev. E **80**, 041110 (2009)
342. A. La Cognata, D. Valenti, A.A. Dubkov, B. Spagnolo, Dynamics of two competing species in the presence of Levy noise sourses. Phys. Rev. E **82**, 011121 (2010)
343. M.D. McDonnell, D. Abbott, What is stochastic resonance? Definitions, microconceptions, debates, and its relevance to biology. PLOS Comput. Biol. **5**, e1000348 (2009)
344. F.Q. Wu, D.J. Menn, X. Wang, Quorum-sensing crosstalk-driven synthetic circuits: From unimodality to trimodality. Chem. Biol. **21**, 1629–1638 (2014)
345. V.A. Slipko, Y.V. Pershin, M. Di Ventra, Changing of the state of a memristive system with white noise. Phys. Rev. E **87**, 041103 (2013)
346. G.A. Patterson, P.I. Fierens, D.F. Grosz, On the beneficial role of noise in resistive switching. Appl. Phys. Lett. **103**, 074102 (2013)
347. S.P. Wen, Z.G. Zeng, T.W. Huang, X.H. Yu, Noise cancellation of memristive neural networks. Neural Netw. **60**, 74–83 (2014)
348. Y. Wang, J. Ma, Y. Xu, F.Q. Wu, P. Zhou, The electrical activity of neurons subject to electromagnetic induction and Gaussian white noise. Int. J. Bifurcation Chaos **27**, 1750030 (2017)
349. D.O. Filatov, D.V. Vrzheshch, O.V. Tabakov, A.S. Novikov, A.I. Belov, I.N. Antonov, V.V. Sharkov, M.N. Koryazhkina, A.N. Mikhaylov, O.N. Gorshkov, A.A. Dubkov, A. Carollo, B. Spagnolo, Noise-induced resistive switching in a memristor-based ZrO2(Y)/Ta2O5 stack. J. Stat. Mechanics Theor. Exp. **2019**, 124026 (2019)
350. F.X. Cai, S. Kumar, T. Van Vaerenbergh, X. Sheng, R. Liu, C. Li, Z. Liu, M. Foltin, S.M. Yu, Q.F. Xia, J.J. Yang, R. Beausoleil, W.D. Lu, J.P. Strachan, Power-efficient combinatorial optimization using intrinsic noise in memristor Hopfield neural networks. Nature Electronics **3**, 409–418 (2020)
351. N.V. Agudov, A.V. Safonov, A.V. Krichigin, A.A. Kharcheva, A.A. Dubkov, D. Valenti, D.V. Guseinov, A.I. Belov, A.N. Mikhaylov, A. Carollo, B. Spagnolo, Nonstationary distributions and relaxation times in a stochastic model of memristor. J. Stat. Mech. Theor. Exp. **2020**, 024003 (2020)

352. A. Stotland, M. Diventra, Stochastic memory: Memory enhancement due to noise. Phys. Rev. E **85**, 011116 (2012)

353. G.A. Patterson, P.I. Fierens, A.A. Garcia, D.F. Grosz, Numerical and experimental study of stochastic resistive switching. Phys. Rev. E **87**, 012128 (2013)

354. A.V. Yakimov, D.O. Filatov, O.N. Gorshkov, D.A. Antonov, D.A. Liskin, I.N. Antonov, A.V. Belyakov, A.V. Klyuev, A. Carollo, B. Spagnolo, Meaurement of the activation energies of oxygen ion diffusion in yttria stabilized zirconia by flicker noise spectroscopy. Appl. Phys. Lett. **114**, 253506 (2019)

355. P.S. Georgiou, I. Koymen, E.M. Drakakis, Noise properties of ideal memristors, 2015 IEEE Int. Symp. Circuits Syst. (ISCAS), IEEE. (2015). https://doi.org/10.1109/ISCAS.2015. 7168841

356. J. Rivnay, P. Leleux, A. Hama, M. Ramuz, M. Huerta, G.G. Malliaras, R.M. Owens, Using white noise to gate organic transistors for dynamic monitoring of cultered cell layers. Sci. Rep. **5**, 11613 (2015)

357. S. Battistoni, R. Sajapin, V. Erokhin, A. Verna, M. Cocuzza, S.L. Marasso, S. Iannotta, Effects of noise sourcing on organic memristive devices. Chaos Solitons Fractals **141**, 110319 (2020)

358. S. Battistoni, V. Erokhin, S. Iannotta, Frequency driven organic memristive devices for neuromorphic short term and long term plasticity. Org. Electronics **65**, 434–438 (2019)

359. T. Ohno, T. Hasegawa, T. Tsuruoka, K. Terabe, J.K. Gimzewski, M. Aono, Short-term plasticity and long-term potentiation mimicked in single inorganic synapse. Nat. Mater. **10**, 591–595 (2011)

360. R.C. Atkinson, R.M. Shiffrin, Human memory: A proposed system and its control process, in *Psychology of Learning and Motivation*, ed. by W. S. Kenneth, S. J. Taylor, (Academic Press, New York, 1968), pp. 89–195

361. P. Gkoupidenis, N. Schaefer, X. Strakosas, J.A. Fairfield, G.G. Malliaras, Synaptic plasticity functions in an organic electrochemical transistor. Appl. Phys. Lett. **107**, 263302 (2016)

362. D. Purves, G.J. Augustine, D. Fitzpatrick, L.C. Katz, A.-S. LaMantia, J.O. McNamara, S.M. Williams, *Neuroscience*, 2nd edn. (Sinauer Associates, Sunderland, 2011)

363. B. Doiron, Y. Zhao, T. Tzounopoulos, Combined LTP and LTD of modulatory inputs controls neuronal processing of primary sensory inputs. J. Neurosci. **31**, 10579–10592 (2011)

364. S. Nabavi, R. Fox, C.D. Proulx, J.Y. Lin, R.Y. Tsien, R. Malinow, Engineering a memory with LTD and LTP. Nature **511**, 348–352 (2014)

365. Z. Wang, S. Joshi, S.E. Savel'ev, H. Jiang, R. Midya, P. Lin, M. Hu, N. Ge, J.P. Strachan, Z. Li, Q. Wu, M. Barnell, G.L. Li, H.L. Xin, R.S. Williams, J.J. Yang, Memristors with diffusive dynamics as synaptic emulators for neuromorphic computing. Nat. Mater. **16**, 101–108 (2017)

366. M.F. Yeckel, A. Kapur, D. Johnston, Multiple forms of LTP in hippocampal CA3 neurons use a common postsynaptic mechanism. Nat. Neurosci. **2**, 625–633 (1999)

367. S. Song, K.D. Miller, L.F. Abbott, Competitive Hebbian learning through spike-timing-dependent synaptic plasticity. Nat. Neurosci. **3**, 919–926 (2000)

368. N. Caporale, Y. Dan, Spike timing-dependent plasticity: A Hebbian learning rule. Annu. Rev. Neurosci. **31**, 25–46 (2008)

369. E.M. Izhikevich, Which model to use for cortical neurons? Neural Netw. **15**, 1063–1070 (2004)

370. S. Ghosh-Dastidar, H. Adeli, Spiking neural networks. Int. J. Neural Syst. **19**, 295–308 (2009)

371. P.A. Merolla, J.V. Arthur, R. Alvarez-Icaza, A.S. Cassidi, J. Sawada, F. Akopyan, B.L. Jackson, N. Imam, C. Guo, Y. Nakamura, B. Brezzo, I. Vo, S.K. Esser, R. Appuswamy, B. Taba, A. Amir, M.D. Flickner, W.P. Risk, R. Manohar, D.S. Modha, A million spiking-neuron integrated circuit with a scalable communication network and interface. Science **345**, 668–673 (2014)

372. J. Schmidhuber, Deep learning in neural networks: An overview. Neural Netw. **61**, 85–117 (2015)

373. C. Zamarreno-Ramos, L.A. Camunas-Mesa, J.A. Perez-Carrasco, T. Masquelier, T. Serrano-Gotarredona, B. Linares-Barranco, On spike-timing-dependent-plasticity, memristive devices, and building a self-learning visual cortex. Front. Neurosci. 5, 26 (2011)

374. Z.Q. Wang, H.Y. Xu, X.H. Li, H. Yu, Y.C. Liu, X.J. Zhu, Synaptic learning and memory functions achieved using oxygen ion migration/diffusion in an amorphous InGaZnO memristor. Adv. Funct. Mater. 22, 2759–2765 (2012)

375. T. Serrano-Gotarredona, T. Masquelier, T. Prodromakis, G. Indiveri, B. Linares-Barranco, STDP and STDP variations with memristors for spiking neuromorphic learning systems. Front. Neurosci. 7, 2 (2013)

376. D.A. Lapkin, A.V. Emelyanov, V.A. Demin, T.S. Berzina, V.V. Erokhin, Spike-timing-dependent plasticity of polyaniline-based memristive element. Microelectron. Eng. 185, 43–47 (2018)

377. D.A. Lapkin, A.V. Emelyanov, V.A. Demin, V.V. Erokhin, L.A. Feigin, P.K. Kashkarov, M.V. Kovalchuk, Polyaniline-based memristive microdevice with high switching rate and endurance. Appl. Phys. Lett. 112, 043302 (2018)

378. P. Gkopidenis, D.A. Koutsouras, G.G. Malliaras, Neuromorphic device architectures with global connectivity through electrolyte gating. Nat. Commun. 8, 15448 (2017)

379. V. Erokhin, T. Berzina, K. Gorshkov, P. Camorani, A. Pucci, L. Ricci, G. Ruggeri, R. Sigala, A. Schuz, Stochastic hybrid 3D matrix: Learning and adaptation of electrical properties. J. Mater. Chem. 22, 22881–22887 (2012)

380. M. Prezioso, M.R. Mahmoodi, F. Merrikh Bayat, H. Nili, H. Vincent, D.B. Strukov, Spike-timing-dependent plasticity learning of coincidence detection with passively integrated memristive circuits. Nat. Commun. 9, 5311 (2018)

381. V. Podzorov, V.M. Pudalov, M.E. Gershenson, Field-effect transistors on rubrene single crystals with parylene gate insulator. Appl. Phys. Lett. 82, 1739–1741 (2003)

382. L.S. Zhou, A. Wanga, S.C. Wu, J. Sun, S. Park, T.N. Jackson, All-organic active matrix flexible display. Appl. Phys. Lett. 88, 083502 (2006)

383. C. Liu, Recent developments in polymer MEMS. Adv. Mater. 19, 3783–3790 (2007)

384. D. Khodagholy, T. Doublet, M. Gurfinkel, P. Quilichini, E. Ismailova, P. Leleux, T. Herve, S. Sanaur, C. Bernard, G.G. Malliaras, Highly conformable conducting polymer electrodes for in vivo recordings. Adv. Mater. 23, H268 (2011)

385. W. Lee, D. Kim, J. Rivnay, N. Matsuhisa, T. Lonjaret, T. Yokota, H. Yawo, M. Sekino, G.G. Malliaras, T. Someya, Integration of organic electrochemical and field-effect transistors for ultraflexible, high temporal resolution electrophysiology arrays. Adv. Mater. 28, 9722 (2016)

386. A. Khiat, S. Cortese, A. Serb, T. Prodromakis, Resistive switching of Pt/TiOx/Pt devices fabricated on flexible parylene-C substrates. Nanotechnology 28, 025202 (2017)

387. B.S. Shvetsov, A.N. Matsukatova, A.A. Minnekhanov, A.A. Nesmelov, B.V. Goncharov, D.A. Lapkin, M.N. Martyshov, P.A. Forsh, V.V. Rylkov, V.A. Demin, A.V. Emelyanov, Parylene based memristors on flexible substrates. Pis'ma v Zhurnal Tekhnicheskoi Fiziki 45, 40–43 (2019)

388. C.T. Cheng, C. Zhu, B.J. Huang, H. Zhang, H.J. Zhang, R. Chen, W.H. Pei, Q. Chen, H.D. Chen, Processing halide perovskite materials with semiconductor technology. Adv. Mater. Technol. 4, 1800729 (2019)

389. D. Ielmini, Resistive switching memories based on metal oxides: Mechanisms, reliability and scaling. Semiconductor Sci. Technol. 31, 063002 (2016)

390. J. Del Valle, J.G. Ramirez, M.J. Rozenberg, I.K. Schuller, Challenges in materials and devices for resistive-switching-based neuromorphic computing. J. Appl. Phys. 124, 211101 (2018)

391. W. Guan, M. Liu, S. Long, Q. Liu, W. Wang, On the resistive switching mechanism of Cu/Zro2:Cu/Pt. Appl. Phys. Lett. 93, 3039079 (2008)

392. A. Bid, A. Bora, A.K. Raychaudhuri, Temperature dependence of the resistance of metallic nanowires of diameter 15 nm: Applicability of Bloch-Gruneisen theory. Phys. Rev. B 74, 035426 (2006)

393. K.E. Nikiruy, A.V. Emelyanov, V.A. Demin, A.V. Sitnikov, A.A. Minnekhanov, V.V. Rylkov, P.K. Kashkarov, M.V. Kovalchuk, Dopamine-like STDP modulation in nanocomposite memristors. AIP Adv. **9**, 065116 (2019)
394. A. Chiolerio, M. Chiappalone, P. Ariano, S. Bocchini, Coupling resistive switching devices with neurons: State of the art and perspectives. Front. Neurosci. **11**, 70 (2017)
395. C. Wang, J. Tang, J. Ma, Minireview on signal exchange between nonlinear circuits and neurons via field coupling. Eur. Phys. J. Special Topics **228**, 1907–1924 (2019)
396. J.S. Tang, F. Yuan, X.K. Shen, Z.R. Wang, M.Y. Rao, Y.Y. He, Y.H. Sun, X.Y. Li, W.B. Zhang, Y.J. Li, B. Gao, H. Qian, G.Q. Bi, S. Song, J.J. Yang, H.Q. Wu, Bridging biological and artificial neural networks with emerging neuromorphic devices: Fundamentals, progress, and challenges. Adv. Mater. **31**, 1902761 (2019)
397. Q.Z. Wan, M.T. Sharbati, J.R. Erickson, Y.H. Du, F. Xiong, Emerging artificial synaptic devices for neuromorphic computing. Adv. Mater. Technol. **4**, 1900037 (2019)
398. A.G. Volkov, C. Tucker, J. Reedus, M.I. Volkova, V.S. Markin, L. Chua, Memristors in plants. Plant Signal. Behav. **9**, e28151 (2014)
399. A.G. Volkov, E.K. Nyasani, C. Tuckett, A.L. Blockmon, J. Reedus, M.I. Volkova, Cyclic voltammetry of apple fruits: Memristors in vivo. Bioelectrochemistry **112**, 9–15 (2016)
400. A.G. Volkov, Biosensors, memristors and actuators in electrical networks of plants. Int. J. Parallel Emerg. Distrib. Syst. **32**, 44–55 (2017)
401. A.G. Volkov, E.K. Nyasani, Sunpatiens compact hot coral: Memristors in flowers. Funct. Plant Biol. **45**, 222–227 (2018)
402. A. Adamatzky, V. Erokhin, M. Grube, T. Schubert, A. Schumann, Physarum chip project: Growing computers from slime mould. Int. J. Unconventional Computing **8**, 319–323 (2012)
403. E. Gale, A. Adamatzky, B. De Lacy Costello, Slime mould memristors. BioNanoScience **5**, 1–8 (2015)
404. A. Cifarelli, T. Berzina, V. Erokhin, Bio-organic memristive device: Polyaniline – Physarum polycephalum interface. Physica Status Solidi C **12**, 218–221 (2015)
405. A. Romeo, A. Dimonte, G. Tarabella, P. D'Angelo, V. Erokhin, S. Iannotta, A bio-inspired memory device based on interfacing Physarum polycephalum with an organic semiconductor. APL Mater. **3**, 014909 (2015)
406. T. Berzina, A. Dimonte, A. Cifarelli, V. Erokhin, Hybrid slime mould-based system for unconventional computing. Int. J. General Systems **44**, 341–353 (2015)
407. G. Tarabella, P. D'Angelo, A. Cifarelli, A. Dimonte, A. Romeo, T. Berzina, V. Erokhin, S. Iannotta, A hybrid living/organic electrochemical transistor based on Physarum polycephalum cell endowed with both sensing and memristive properties. Chem. Sci. **6**, 2859–2868 (2015)
408. E. Braund, E.R. Miranda, On the building practical biocomputers for real-world applications: Receptacles for culting slime mould memristors and component standardization. J. Bionic Engineering **14**, 151–162 (2017)
409. E.R. Miranda, E. Braund, A method for growing bio-memristors from slime mold. JOVE J. Visual. Exp. **129**, e5676 (2017)
410. L. Wang, D. Wen, Nonvolatile bio-memristor based on silkworm hemolymph proteins. Sci. Rep. **7**, 17418 (2017)
411. O. Pabst, O.G. Martinsen, L. Chua, The non-linear electrical properties of human skin make it a generic memristor. Sci. Rep. **8**, 15806 (2018)
412. A. Serb, A. Corna, R. George, A. Khiat, F. Rocchi, M. Reato, M. Maschietto, C. Mayr, G. Indiveri, S. Vassanelli, T. Prodromakis, Memristive synapses connect brain and silicon spiking neurons. Sci. Rep. **10**, 2590 (2020)
413. S.T. Keene, C. Lubrano, S. Kazemzadeh, A. Melianes, Y. Tuchman, G. Polino, P. Scognamiglio, L. Cina, A. Salleo, Y. Van de Burg, F. Santoro, A biohybrid synapse with neurotransmitter-mediated plasticity. Nat. Mater. **19**, 969–973 (2020)

414. E. Juzekaeva, A. Nasretdinov, S. Battistoni, T. Berzina, S. Iannotta, R. Khazipov, V. Erokhin, M. Mukhtarov, Coupling cortical neurons through electronic memristive synapse. Adv. Mater. Technol. **4**, 1800350 (2019)

415. P. Fromherz, A. Offenhausser, T. Vetter, J. Weis, A neuron-silicon junction: A Retzius cell of the leech on an insulated-gate field-effect transistor. Science **252**, 1290–1293 (1991)

416. S. Vassanelli, P. Fromherz, Transistor records of excitable neurons from rat brain. Appl. Phys. A **66**, 459–463 (1998)

417. D. Kuzum, S. Yu, H.-S.P. Wong, Synaptic electronics: Materials, devices and applications. Nanotechnology **24**, 382001 (2013)

418. C. Rossant, S.N. Kadir, D.F.M. Goodman, J. Schulman, M.L.D. Hunter, A.B. Saleem, A. Grosmark, M. Belluscio, G.H. Denfield, A.S. Ecker, A.S. Tolias, S. Solomon, G. Buzsaki, M. Carandini, K.D. Harris, Spike sorting for large, dense electrode arrays. Nat. Neurosci. **19**, 634–641 (2016)

419. G. Buzsaki, E. Stark, A. Berenyi, D. Khodagholy, D.R. Kipke, E. Yoon, K.D. Wise, Tools for probing local circuits: High-density silicon probs combined with optogenetics. Neuron **86**, 92–105 (2015)

420. S.J. Etherington, S.E. Atkinson, G.J. Stuart, S.R. Williams, *Synaptic Integration in eLS* (John Wiley and Sons Ltd., New Jersey, 2010)

421. D. Fricker, R. Miles, EPSP amplification and the precision of spike timing in hipposampal neurons. Neuron **28**, 559–569 (2000)

422. G. Buzsaki, A. Draguhn, Neuronal oscillations in cortical networks. Science **304**, 1926–1929 (2004)

423. B.P. Timko, T. Cohen-Karni, Q. Qing, B. Tian, C.M. Lieber, Design and implementation of functional nanoelectronics interfaces with biomolecules, cells, and tissue using nanowire device arrays. IEEE Trans. Nanotechnology **9**, 269–280 (2009)

424. R.R. Harrison, C. Charles, A low-power low-noise CMOS amplifier for neural recording applications. IEEE J. Solid State Circuits **38**, 958–965 (2003)

425. F. Patolsky, B.P. Timko, G. Yu, Y. Fang, A.B. Greytak, G. Zheng, C.M. Lieber, Detection, simulation, and inhibition of neuronal signals with high-density nanowire transistor arrays. Science **313**, 1100–1104 (2006)

426. D. Khodagholy, J.N. Gelinas, T. Thesen, W. Doyle, O. Devinsky, G.G. Malliaras, G. Buzsaki, NeuroGrid: Recording action potentials from the surface of the brain. Nat. Neurosci. **18**, 310–315 (2015)

427. D. Khodagholy, T. Doublet, P. Quilichini, M. Gurfinkel, P. Leleux, A. Ghestem, E. Ismailova, T. Herve, S. Sanaur, C. Bernard, G.G. Malliaras, In vivo recording of brain activity using organic transistors. Nat. Commun. **4**, 1575 (2013)

428. M.C. Peterman, J. Noolandi, M.S. Blumenkranz, H.A. Fishman, Localized chemical release from an artificial synapse chip. Proc. Natl. Acad. Sci. U. S. A. **101**, 9951–9954 (2004)

429. C.M. Rountree, A. Raghunathan, J.B. Troy, L. Saggere, Prototype chemical synapse chip for spatially patterned neurotransmitter stimulation of the retina ex vivo. Microsyst. Nanoeng. **3**, 17052 (2017)

430. M.C. Peterman, N.Z. Mehenti, K.V. Bilbao, C.J. Lee, T. Leng, J. Noolandi, S.F. Bent, M.S. Blumenkranz, H.A. Fishman, The artificial synapse chip: A flexible retinal interface based on directed retinal cell growth and neurotransmitter simulation. Artif. Organs **27**, 975–985 (2003)

431. F. Alibart, E. Zamanidoost, D.B. Strukov, Pattern classification by memristive crossbar circuits using ex situ and in situ training. Nat. Commun. **4**, 2072 (2013)

432. L.J. Juarez-Hernandez, N. Cornella, L. Pasquardini, S. Battistoni, L. Vidalino, L. Vanzetti, S. Caponi, M. Dalla Serra, S. Iannotta, C. Pederzolli, P. Macchi, C. Musio, Bio-hybrid interfaces to study neuromorphic functionalities: New multidisciplinary evidences of cell viability on poly(aniline)(PANI), a semiconductor polymer with memristive properties. Biophys. Chem. **208**, 40–47 (2016)

433. R.A. Kaul, N.I. Syed, P. Fromherz, Neuron-semiconductor chip with chemical synapse between identified neurons. Phys. Rev. Lett. **92**, 038102 (2004)

434. A. Dimonte, T. Berzina, A. Cifarelli, V. Chiesi, F. Albertini, V. Erokhin, Conductivity patterning with Physarum polycephalum: Natural growth and deflecting. Physica Status Solidi C **12**, 197–201 (2015)

435. S. Erokhina, V. Sorokin, V. Erokhin, Skeleton-supported stochastic networks of organic memristive devices: Adaptation and learning. AIP Adv. **5**, 027129 (2015)

436. C. Auner, U. Palfinger, H. Gold, J. Kraxner, A. Haase, T. Haber, M. Sezen, W. Grogger, G. Jakopic, J.R. Krenn, G. Leising, B. Stadlober, High-performing submicron organic thin-film transistors babricated by residue-free embossing. Org. Electronics **11**, 552–557 (2010)

437. M.V.L. Bennett, R.S. Zukin, Electrical coupling and neuronal synchronization in the mammalian brain. Neuron **19**, 495–510 (2004)

438. M. Talanov, I. Lavrov, C. Menshenin, Comptational modeling of spinal locomotor circuitry. Eur. J. Clin. Invest. **48**, 224 (2018)

439. M. Talanov, I. Lavrov, V. Erokhin, Modelling reflex arc for a memristive implementation. Eur. J. Clin. Invest. **48**, 223 (2018)

440. D.B. Strukov, R.S. Williams, Four-dimensional address topology for circuits with stacked multilayer crossbar arrays. Proc. Natl. Acad. Sci. U. S. A. **106**, 20155–20158 (2009)

441. S. Schweiger, M. Kubicek, F. Messerschmitt, C. Murer, J.L.M. Rupp, A microdot multilayer oxide device: Let us tune the strain-ionic transport interaction. ACS Nano **8**, 5032–5048 (2014)

442. X. Hu, G. Feng, S. Duan, L. Liu, Multilayer RTD-memristor-based cellular neural networks for color image processing. Neurocomputing **162**, 150–162 (2015)

443. I. Michelakaki, P. Bousoulas, N. Maragos, N. Boukos, D. Tsoukalas, Resistive memory multilayer structure with self-rectifying and forming free midlayer. J. Vacuum Sci. Technol. A **35**, 021501 (2017)

444. R. Hasan, T.M. Taha, C. Yakopcic, On-chip training of memristor crossbar based multi-layer neural networks. Microelectronics J. **66**, 31–40 (2017)

445. S.G. Zhang, Fabrication of novel biomaterials through molecular self-assembly. Nat. Biotechnol. **21**, 1171–1178 (2003)

446. J.Y. Chen, A.M. Mayes, C.A. Ross, Nanostructure engineering by templated self-assembly of block copolymers. Nat. Mater. **3**, 823–828 (2004)

447. P.W.K. Rothemund, Folding DNA to create nanoscale shapes and pattern. Nature **440**, 297–302 (2006)

448. J.Y. Cheng, C.A. Ross, H.I. Smith, E.L. Thomas, Templated self-assembly of block copolymers: Top-down helps bottom-up. Adv. Mater. **18**, 2505–2521 (2006)

449. K. Ariga, Y. Yamauchi, G. Rydzek, Q. Ji, Y. Yonamine, K.C.W. Wu, J.P. Hill, Layer-by-layer nanoarchitectronics: Invention, innovation, and evolution. Chem. Lett. **43**, 36–68 (2014)

450. V. Erokhin, T. Berzina, P. Camorani, M.P. Fontana, Conducting polymer – solid electrolyte fibrillar composite material for adaptive networks. Soft Matter **2**, 870–874 (2006)

451. V. Erokhin, Polymer-based adaptive networks, in *The New Frontiers of Organic and Composite Nanotechnologies*, ed. by V. Erokhin, M. K. Ram, O. Yavuz, (Elsevier, Oxford, Amsterdam, 2008), pp. 287–353

452. A. Cifarelli, A. Dimonte, T. Berzina, V. Erokhin, Non-linear bioelectronic element: Schottky effect and electrochemistry. Int. J. Unconven. Comput. **10**, 375–379 (2014)

453. T. Berzina, A. Pucci, G. Ruggeri, V. Erokhin, M.P. Fontana, Gold nanoparticles – polyaniline composite material: Synthesis, structure and electrical properties. Synth. Met. **161**, 1408–1413 (2011)

454. A. Pucci, N. Tirelli, E.A. Willneff, S.L.M. Schoeder, F. Galembeck, G. Ruggeri, Evidence and use of metal-chromophore interactions: Luminescence dichroism of terthiophene-coated gold nanoparticles in polyethylene oriented films. J. Mater. Chem. **14**, 3495–3502 (2004)

455. K.G. Thomas, P.V. Kamat, Chromophore-functionalized gold nanoparticles. Acc. Chem. Res. **36**, 888–898 (2003)

456. Z. Atay, T. Biver, A. Corti, N. Eltugral, E. Lorenzini, M. Masini, A. Paolicchi, A. Pucci, G. Ruggeri, F. Secco, M. Venturini, Non-covalent interactions of cadmium sulphide and gold nanoparticles with DNA. J. Nanopart. Res. **12**, 2241–2253 (2010)

457. I. Gofberg, D. Mandler, Preparation and comparison between different thiol-protected Au nanoparticles. J. Nanopart. Res. **12**, 1807–1811 (2010)

458. F. Alibart, S. Pleutin, D. Guerin, C. Novembre, S. Lenfant, K. Lmimouni, C. Gamrat, D. Vuillaume, An organic nanoparticle transistor behaving as a biological spiking synapse. Adv. Funct. Mater. **20**, 330–337 (2010)

459. V. Erokhin, On the learning of stochastic networks of organic memristive devices. Int. J. Unconventional Computing **9**, 303–310 (2013)

460. J.W. Lichtman, H. Colman, Synapse elimination and indelible memory. Neuron **25**, 269–278 (2000)

461. G.M. Innocenti, D.O. Frost, The postnatal development of visual callosal connections in the absence of visual experience or of the eyes. Exp. Brain Res. **39**, 365–375 (1980)

462. D. Purves, J.W. Lichtman, Elimination of synapses in the developing nervous system. Science **210**, 153–157 (1980)

463. R. Apfelbach, E. Weiler, Olfactory depriviation enhances normal spine loss in the olfactory bulb of developing ferrets. Neurosci. Lett. **62**, 169–173 (1985)

464. G. Buzsaki, *Rhythms of the Brain* (Oxford University Press, Oxford, 2006)

465. R. Sigala, A. Smerieri, A. Schuz, P. Camorani, V. Erokhin, Modeling and simulating the adaptive electrical properties of stochastic polymeric 3D networks. Model. Simul. Mater. Sci. Eng. **21**, 075007 (2013)

466. V. Litovski, M. Zwolinski, *VLSI Circuit Simulation and Optimization* (Chapman and Hall, London, 1997)

467. D.J. Watts, S.H. Strogatz, Collective dynamics of "small-world" networks. Nature **393**, 440–442 (1998)

468. A.L. Barabasi, R. Albert, Emergence of scaling in random networks. Science **286**, 509–512 (1999)

469. V. Braitenberg, *Cell Assemblies in the Cerebral Cortex. Theoretical Approaches to Complex Systems* (Springer, Berlin, 1978), pp. 171–188

470. B. Hellwig, A. Schuz, A. Aertsen, Synapses on axon collaterals of pyramidal cells are spaced at random intervals: A golgi study in the mouse cerebral cortex. Biol. Cybern. **71**, 1–12 (1994)

471. M.B. Young, J.W. Scannell, G. Burns, *The Analysis of Cortical Connectivity* (Springer, New York, 1995)

472. A. Schuz, V. Braitenberg, The human cortical white matter: Quantitative aspects of cortico-cortical long-range connectivity, in *Cortical Areas: Unity and Diversity*, ed. by A. Schuz, R. Miller, (Taylor and Francis, London, 2002), pp. 377–386

473. M. Ismail, H. Abbas, C. Choi, S. Kim, Stabilized and RESET-voltage controlled multi-level switching characteristics in ZrO2-based memristors by inserting a-ZTO interface layer. J. Alloys Compd. **835**, 155256 (2020)

474. S. Goswami, A.J. Matula, S.P. Rath, S. Hedstrom, S. Saha, M. Annamalai, D. Sengupta, A. Patra, S. Ghosh, H. Jani, S. Sarkar, M.R. Motapothula, C.A. Nijhuis, J. Martin, S. Goswami, V.S. Batista, T. Venkatesan, Robust resistive memory devices using solution-processable metal-coordinated azo aromatics. Nat. Mater. **16**, 1216 (2017)

475. A. Mehonic, S. Cueff, M. Wojdak, S. Hudziak, O. Jambois, C. Labbe, B. Garrido, R. Rizk, A.J. Kenyon, Resistive switching in silicon suboxide films. J. Appl. Phys. **111**, 074507 (2012)

476. C.B. Pan, Y.F. Ji, N. Xiao, F. Hui, K.C. Tang, Y.Z. Guo, X.M. Xie, F.M. Puglisi, L. Larcher, E. Miranda, L.L. Jiang, Y.Y. Shi, I. Valov, P. McIntyre, R. Waser, M. Lanza, Coexistence of grain-boundaries-assisted bipolar and threshold resistive switching in multilayer hexagonal boron nitride. Adv. Funct. Mater. **27**, 1604811 (2017)

477. Y.N. Malakhova, A.N. Korovin, D.A. Lapkin, S.N. Malakhov, V.V. Shcherban, E.B. Pichkur, S.N. Yakunin, V.A. Demin, S.N. Chvalun, V. Erokhin, Planar and 3D fibrous polyaniline-based materials for memristive elements. Soft Matter **13**, 7300–7306 (2017)

478. C. Du, M.A. Zidan, W. Ma, S.H. Lee, W.D. Lu, Reservoir computing using dynamic memristors for temporal information processing. Nat. Commun. **8**, 2204 (2017)

479. G. Tanaka, T. Yamane, J.B. Heroux, R. Nakane, N. Kanazawa, S. Takeda, H. Numata, D. Nakano, A. Hirose, Recent advances in physical reservoir computing: A review. Neural Netw. **115**, 100–123 (2019)

480. V. Erokhin, Memristive devices for neuromorphic applications: Comparative analysis. BioNanoScience **10**, 834–847 (2020)

481. A. Gelencser, T. Prodromakis, C. Toumazou, T. Roska, Biomimetic model of the outer plexiform by incorporating memristive devices. Phys. Rev. E **85**, 041918 (2012)

482. P.M. Sheridan, F.X. Cai, C. Du, W. Ma, Z.Y. Zhang, W.D. Lu, Sparse coding with memristor networks. Nat. Nanotechnol. **12**, 784 (2017)

483. L. Bao, J. Kang, Y.C. Fang, Z.Z. Yu, Z.W. Wang, Y.C. Yang, Y.M. Cai, R. Huang, Artificial shape perception retina network based on tunable memristive neurons. Sci. Rep. **8**, 13727 (2018)

484. X. Ji, X.F. Hu, Y. Zhou, Z.K. Dong, S.K. Duan, Adaptive sparse coding based on memristive neural network with applications. Cogn. Neurodyn. **13**, 475–488 (2019)

485. G.-Q. Min, L.-D. Wang, S.-K. Duan, Chaotic circuit of ion migration memristor and its application in the voice secure communication. *Acta Phys. Sinica* **64**, 210507 (2015)

486. P. Liu, R. Xi, P.B. Ren, J.L. Hou, X. Li, Analysis and implementation of a new switching memristor Scroll hyperchaotic system and application in secure communication. Complexity **2018**, 3497640 (2018)

487. S.H. Sung, D.H. Kim, T.J. Kim, I.S. Kang, K.J. Lee, Unconventional inorganic-based memristive devices for advanced intelligent systems. Adv. Mater. Technol. **4**, 1900080 (2019)

488. K. Rajagopal, S. Kacar, Z.C. Wei, P. Duraisamy, T. Kifle, A. Karthikeyan, Dynamical investigation and chaotic associated behaviors of memristor Chua's circuit with a non-ideal voltage-controlled memristor and its application to voice encryption. AEU Int. J. Electronic Commun. **107**, 183–191 (2019)

489. S. Vaidyanathan, A.T. Azar, A. Akgul, C.H. Lien, S. Kacar, A memristor-based system with hidden hyperchaotic attractors, its circuit design, synchronization via integral sliding mode control and an application to voice encryption. Int. J. Automation Control. **13**, 644–667 (2019)

490. Z. Nagy, P. Szolgay, Configurable multilayer CNN-UM emulator on FPGA. IEEE Trans. Circuits and Systems I **50**, 774–778 (2003)

491. Y.H. Sun, X. Zheng, X.Q. Yan, Q.L. Liao, S. Liu, G.J. Zhang, Y. Li, Y. Zhang, Bioinspired tribotronic resistive switching memory for self-powered memorizing mechanical stimuli. ACS Appl. Mater. Interfaces **9**, 43822–43829 (2017)

492. Z. Wang, Q. Hong, X. Wang, Memristive circuit design of emotional generation and evolution based on skin-like sensory processor. IEEE Trans. Biomed. Circuits Syst. **13**, 631–644 (2019)

493. C. Zhang, W.B. Ye, K. Zhou, H.Y. Chen, J.Q. Yang, G.L. Ding, X.L. Chen, Y. Zhou, L. Zhou, F.J. Li, S.T. Han, Bioinspired artificial sensory nerve based on nafion memristor. Adv. Funct. Mater. **29**, 1808783 (2019)

494. R. Kozma, M. Puljic, Hierarchial random cellular neural networks for system-level brain-like signal processing. Neural Netw. **45**, 101–110 (2013)

495. F.C. Zhou, J.W. Chen, X.M. Tao, X.R. Wang, Y. Chai, 2D materials based optoelectronic memory: Convergence of electronic memory and optical sensor. Research **2019**, 9490413 (2019)

496. S. Carrara, D. Sacchetto, M.A. Doucey, C. Baj-Rossi, G. De Micheli, Y. Leblebici, Memristive biosensors: A new detection method by using nanofabricated memristors. Sens. Actuators B **171**, 449–457 (2012)

497. M.-A. Doucey, S. Carrara, Nanowire sensors in cancer. Trends Biotechnol. **37**, 86–99 (2019)

498. D.P. Sahu, S.N. Jammalamadaka, Detection of bovine serum albumin using hybrid $TiO2$ + graphene oxide baised bio-resistive random access memory device. Sci. Rep. **9**, 16141 (2019)

499. N.S.M. Hadis, A. Abd Manaf, S.H. Herman, Trends of deposition and patterning techniques of TiO2 for memristor based bio-sensing applications. Microsyst. Technol. Micro Nanosyst. Inf. Storage Process. Syst. **19**, 1889–1896 (2013)

500. A. Adamatzky, Slime mould processors, logic gates and sensors. Philos. Trans. Royal. Soc. A **373**, 20140216 (2014)

501. J.C. Shank, M.B. Tellekamp, W.A. Doolittle, Evidence of ion intercalation mediated band structure modification and opto-ionic coupling in lithium niobite. J. Appl. Phys. **117**, 035704 (2015)

502. C. Nyenke, L. Dong, Fabrication of a W/CuxO/Cu memristor with sub-micron holes for passive sensing of oxygen. Microelectron. Eng. **164**, 48–52 (2016)

503. A.A. Haidry, A. Ebach-Stahl, B. Saruhan, Effect of Pt/TiO2 interface on room temperature hydrogen sensing performance of memristor type Pt/TiO2/Pt structure. Sens. Actuators B **253**, 1043–1054 (2017)

504. S. Choi, S. Kim, J. Jang, G. Ahn, J.T. Jang, J. Yoon, T.J. Park, B.G. Park, D.M. Kim, S.J. Choi, S.M. Lee, E.Y. Kim, H.S. Mo, D.H. Kim, Implementing and artificial synapse and neuron using a Si nanowire ion-sensitive field-effect transistor and indium-gallium-zinc-oxide memristors. Sens. Actuators B **296**, 126616 (2019)

505. M. Vidis, T. Plecenik, M. Mosko, S. Tomasec, T. Roch, L. Strapinskyy, B. Grancic, A. Plecenik, Gasistor: A memristor based gas-triggered switch and gas sensor with memory. Appl. Phys. Lett. **115**, 093504 (2019)

506. L. Lunelli, C. Collini, A.M. Jimenez-Garduno, A. Roncador, G. Giusti, R. Verucchi, L. Pasquardini, S. Iannotta, P. Macchi, L. Lorenzelli, C. Pederzolli, C. Musio, C. Potrich, Prototyping a memristive-based device to analyze neuronal excitability. Biophys. Chem. **253**, 106212 (2019)

507. G.A. Illarionov, D.S. Kolchanov, O.A. Zhukov, E. Sergeeva, V.V. Krishtop, A.V. Vinogradov, M.I. Morozov, Inkjet assisted fabrication of planar biocompatible memristors. RSC Adv. **9**, 35998–36004 (2019)

508. Y.M. Qi, B. Sun, G.Q. Fu, T.T. Li, S.H. Zhu, L. Zheng, S.S. Mao, X. Kan, M. Lei, Y.Z. Chen, A nonvolatile organic resistive switching memory based on lotus leaves. Chem. Phys. **516**, 168–174 (2019)

509. L. Wang, D. Wen, Resistive switching memory devices based on body fluid of Bombyx mori L. Micromachines **10**, 540 (2019)

510. A. Mikhaylov, A. Pimashkin, Y. Pigareva, S. Gerasimova, E. Gryaznov, S. Shchanikov, A. Zuev, M. Talanov, I. Lavrov, V. Demin, V. Erokhin, S. Lobov, I. Mukhina, V. Kazantsev, H. Wu, B. Spagnolo, Neurohybrid memristive CMOS-integrated systems for biosensors and neuroprosthesis. Front. Neurosci. **14**, 358 (2020)

511. S.G. Kim, J.S. Han, H. Kim, S.Y. Kim, H.W. Jang, Recent advances in memristive materials for artificial synapses. Adv. Mater. Technol. **3**, 1800457 (2018)

512. V.A. Slipko, Y.V. Pershin, M. Di Ventra, Changing the state of a memristive system with white noise. Phys. Rev. E **87**, 042103 (2013)

513. L.Y. Wang, J. Yang, Y. Zhu, M.D. Yi, L.H. Xie, R.L. Ju, Z.Y. Wang, L.T. Liu, T.F. Li, C.X. Zhang, Y. Chen, Y.N. Wu, W. Huang, Rectification-regulated memristive characteristics in electron-type CuPc-based element for electrical synapse. Adv. Electron. Mater. **3**, 1700063 (2017)

514. L.Y. Shi, G.H. Zheng, B.B. Tian, B. Dkhil, C.G. Duan, Research progress on solutions to the sneak path issue in memristor crossbar arrays. Nanoscale Adv. **2**, 1811–1827 (2020)

515. S.K. Sharma, D. Haobijam, S.S. Singh, M.Z. Malik, R.K.B. Singh, Neuronal communication: Stochastic neuron dynamics and multi-synchrony states. AEU Int. J. Electron. Commun. **100**, 75–85 (2019)

516. A. Damasio, *The Feeling of What Happens: Body and Emotion in the making of consciousness* (Harcourt Inc., New York, 1999)

517. A. Ortony, G.L. Clore, A. Collins, *The cognitive structure of emotions* (Cambridge University Press, Cambridge, 1990)

518. H. Lovheim, A new three-dimensional model for emotions and monoamine neuro-transmitters. Med. Hypotheses **78**, 341–348 (2012)
519. M. Talanov, A. Toschev, A. Leukhin, Modeling the fear-like state in realistic neural networks. BioNanoScience **7**, 446–448 (2017)
520. J. Vallverdu, M. Talanov, S. Distefano, M. Mazzara, A. Tchitching, I. Nurgaliev, A cognitive architecture for the implementation of emotions in computing systems. Biol. Inspired Cogn. Arch. **15**, 34–40 (2016)
521. J. Vallverdu, G. Trovato, Emotional affordances for human-robot interaction. Adapt. Behav. **24**, 320–334 (2016)

Index

© Springer Nature Switzerland AG 2022
V. Erokhin, *Fundamentals of Organic Neuromorphic Systems*,
https://doi.org/10.1007/978-3-030-79492-7

Printed in the United States
by Baker & Taylor Publisher Services